Irrigation Technology
Theory and Practice

Irrigation Technology

Theory and Practice

Amalendu Bhattacharya

RANDOM

Irrigation Technology: Theory and Practice

ISBN 978-93-5111-366-9

Published in 2014 in India by

RANDOM PUBLICATIONS

4376-A/4B, Gali Murari Lal, Ansari Road
New Delhi-110 002
Phone : +91-11-43580356, +91-11-23289044
e-mail: randomexports@gmail.com, sales@randompublications.com, info@randompublications.com

Reprinted 2022

Type Setting by: Friends Media, Delhi-110089
Digitally Printed at: Replika Press Pvt. Ltd.

Preface

Various types of irrigation techniques differ in how the water obtained from the source is distributed within the field. In general, the goal is to supply the entire field uniformly with water, so that each plant has the amount of water it needs, neither too much nor too little. The modern methods are efficientenough to achieve this goal. In surface irrigation systems, water moves over and across the land by simple gravity flow in order to wet it and to infiltrate into the soil. Surface irrigation can be subdivided into furrow, borderstrip or basin irrigation. It is often called flood irrigation when the irrigation results in flooding or near flooding of the cultivated land. Historically, this has been the most common method of irrigating agricultural land. Where water levels from the irrigation source permit, the levels are controlled by dikes, usually plugged by soil. This is often seen in terraced rice fields (rice paddies), where the method is used to flood or control the level of water in each distinct field. In some cases, the water is pumped, or lifted by human or animal power to the level of the land. Drip irrigation, also known as trickle irrigation, functions as its name suggests. In this system water falls drop by drop just at the position of roots. Water is delivered at or near the root zone of plants, drop by drop. This method can be the most water-efficient method of irrigation, if managed properly, since evaporation and runoff are minimized.

In modern agriculture, drip irrigation is often combined with plastic mulch, further reducing evaporation, and is also the means of delivery of fertilizer. The process is known as fertigation. Deep percolation, where water moves below the root zone, can occur if a drip system is operated for too long or if the delivery rate is too high. Drip irrigation methods range from very high-tech and computerized to low-tech and labour-intensive. Lower water pressures are usually needed than for most other types of systems, with the exception of low energy centre pivot systems and surface irrigation systems, and the system can be designed for uniformity throughout a field or for precise water delivery to individual plants in a landscape containing a mix of plant species. Although it is difficult to regulate pressure on steep slopes, pressure compensating emitters are available, so the field does not have to be level. High-tech solutions involve precisely calibrated emitters located along lines of tubing that extend from a computerized set of valves. In sprinkler or overhead irrigation, water is piped to one or more central locations within the field and distributed by overhead high-pressure sprinklers or guns. A system utilizing sprinklers, sprays, or guns mounted overhead on permanently installed risers is often referred to as asolid-set irrigation system. Higher pressure sprinklers that rotate are called rotors and are driven by a ball drive, gear drive, or impact mechanism. Sources of irrigation water can be groundwater extracted from springs or by using wells, surface water withdrawn from rivers, lakes or reservoirs or non-conventional sources like treated wastewater, desalinated water or drainage water. A special form of irrigation using surface water is spate irrigation, also called floodwater harvesting. In case of a flood (spate) water is diverted to normally dry river beds (wadis) using a network of dams, gates and channels and spread over large areas. The moisture stored in the soil will be used thereafter to grow crops. Spate irrigation areas are in particular located in semi-arid or arid, mountainous regions. While floodwater harvesting belongs to the accepted irrigation methods, rainwater harvesting is usually not considered as a form of irrigation. Rainwater harvesting is the collection of runoff water from roofs or unused land and the concentration.

It is expected that the book will provide a fund of rich experiences to the students and teachers.

—Amalendu Bhattacharya

Contents

1

Water Resources and Distribution

It is appropriate to examine the concept of water resource systems as a foundation for consideration of the distribution issues that impact human use of the resource throughout the world. One helpful approach is to consider a basic problem in the provision of water, namely the disparity between the distribution of available water versus the demand for the resource in a specific region or locale (Buras, 1972). There are three basic types of maldistribution of water resources. First, one may envision a geographical/spatial maldistribution of the resource; the geographical area may be experiencing a flood condition (excess water) or a drought condition (insufficient water). In either case, the water availability fails to match the present demand. A second type of maldistribution of water is temporal. For example, the water may be available during the spring of the year following snowmelt in upstream areas. However, this water may not be available later in the growing season when it is needed for irrigation purp oses.

The third type of maldistribution of water as a resource is qualitative. The mineral, biological, and physical attributes of the water as it is available at specific locations may limit the use of the water for beneficial uses. For example, brackish water may be unfit for human and animal consumption and may have limited value for other purposes. One role for the field of water resource systems is to undertake the studies and research needed to enable sufficient water to be available within a specific locality or region in order to meet both present as well as future demands. Traditionally, it has been the task of the water resource professionals to apply techniques of systems analysis to this type of maldistribution problem and to identify ways to overcome the maldistribution of water and to mitigate, correct, and provide the resource on an as needed basis. This approach is still valid; however, it needs to be broadened to consider ways to alter the demand side of the water need equation as well as meeting the supply side of the water need equation.

Systems Analysis Methods: Multi-objective Planning

Normally one may expect a series of inter-related steps to be undertaken to accomplish a systems analysis of a problem of interest. Clearly, the methods of systems analysis may be applied to a wide spectrum of problems including but not limited to water resource problems and issues. The first step is to identify and quantify the objectives associated with the problem. For example, one objective may be to provide water of a specified quality and quantity to a specific region with a specified degree of reliability. A second related objective might be to provide this desired level of water service at minimum economic cost.

Any problem that contains two or more objectives is considered a multi-objective problem. The solutions to this class of problem result in the identification of a range of potential solutions where the gain of one objective, for example, lower economic cost is achieved only through the tradeoff against other objectives — i.e., less reliable water service in the example. Included in the probl em definition portion of the systems analysis is the definition of decision variables, for example, the location and the level of treatment provided to achieve desired levels of water quality for the service area. In addition, the analysis requires the formation of constraints that describe the physical limitations associated with the problem at hand. The objective functions and the constraint equations which include the decision variables for the problem provide the analytical framework for the systems analysis to be undertaken.

A second step in the overall process is to collect data relevant to the problem. This data may be physical data, biological data, and/or chemical data that characterize the water resource in its present condition. Economic data including cost data is obtained which specifies information for this example on the costs associated with providing the desired level and reliability of water service. Simulation models and other techniques also may be utilized in order to calculate values of coefficients associated with the decision variables for the multi-objective optimization problem. Having assembled the necessary data and identified the objective functions and associated constraint equations, the third step is to generate alternatives to achieve the desired objectives. In carrying out this third step in the overall systems analysis, one is identifying potential solutions in objective function space, which satisfy all of the constraints in the problem and identify optimum solution points for the objective functions in the problem. For example, if cost is to be minimized one may set different levels of water service to be provided and then find the least cost solution to provide that specified level of water service.

The fourth step is to evaluate trade-offs between the objective functions that are being considered in the problem at hand. It is very important for the decision makers to have this information before them so that it is clear what increased risk they may incur if they choose to implement lower cost solutions for the provision of water service. It is this explicit trade-off between objectives that is a critical and important element in multi-objective planning for water resource development and use. This fourth step leads directly to the fifth step, namely the selection by the decision-makers of a preferred alternative from among the large number of potential solutions to the problem.

The sixth and final step is the implementation and continued maintenance of the preferred and selected alternative. This step may be the most complex of all of the steps in the systems analysis process. The implementation step requires the commitment of sufficient financial resources to build and to operate the water project. The ability to secure and maintain these financial resources often requires a major political commitment from the multiple units of government served by the project. It should be clear from this brief presentation that the planning and management of water resources is a demanding task requiring the contribution of knowledge and professional skills of many disciplines including planners, engineers, policy makers, economists and others.

Major Planning Issues: Resolution Needed

In undertaking a multi-objective planning activity in the field of water resources, there are a number of major issues which need to be addressed and resolved in order for the planning to proceed in an effective way. First of these issues is to decide upon the desired scale of development For example, is this water project to serve a limited locale or is it to serve a whole region or a whole watershed. Closely related to the scale of development is the sizing of the physical structural elements to be included in the project. This issue relates not only to the scale of development but also to the desired reliability to be achieved once the project is completed. For example, if the project is designed to protect human life, the reliability or safety factor associated with the project may require larger structures to provide the extra degree of protection of human life. If the water project involves certain control structures such as reservoirs then another issue that must be addressed is to develop a desired o perating policy for these control structures so that they can perform their function with a high probability of success. A further dimension to be considered in the planning of major water projects is to minimize undesirable environmental impacts associated with the implementation and operation of the projects. Finally, it is essential to provide for the establishment of an adequate organizational structure to implement and maintain the water resource project. This organizational structure needs to have the capacity to ensure the physical, financial, operational, technical, political, social, and environmental integrity of the project. This issue is among the most complex and challenging of all the tasks facing the water resource planners.

There are three major important issue areas that need to be considered to gain perspectives on global water problems facing the world as we move into the 21st century. These areas will be discussed and data will be presented to illustrate the impacts of these problems in certain areas of the world. The first key issue is the growth of global population against a relatively fixed supply of fresh water. The hard fact is simply that population growth is a driver that acts to increase the use of water to meet basic human needs. In addition to meeting the immediate individual need for water to sustain the individual, vast quantities of water are consumed through irrigation for the production of food and fiber to meet human needs. Additional large quantities of water are utilized for cooling purposes in the generation of fossil and nuclear energy. Given that the fresh water resources are not distributed uniformly across the globe, the increasing demand for the essential water resources requires that the water be u sed as efficiently as possible. Furthermore, it requires that means need to be found that will ensure the reliable provision of adequate food/fiber for those people living in water scarce regions.

A second key issue is the uncertainty associated with climate change and its potential impact upon the availability of fresh water resources in specific regions of the world. The risk of reducing the quantity of fresh water is of particular concern in regions that are already short of water needed to meet present needs. There can be increased evaporation as a consequence of global warming, soil moisture may decrease because of increased evaporation, and snowfall/snowmelt patterns may change. It is clear that the future will not be like the past; however, it is not clear what the future will bring other than change.

The third major global issue is that of sustainability and water resources. This topic is complex and not fully defined. The issue of population growth as a driver has already been discussed; it is important to attach a second driver to population growth and this second element is the standard of living. As the standard of living increases, increased demands are placed upon regional water resources to support the demands of people to achieve higher and higher standards of living. There are various technical means that may be utilized to augment existing water supplies. Desalination is very energy intensive and is applied in very special circumstances. Large scale, long distance transport of water has been practiced for many years to capture and move water from relatively water rich areas to areas where the demand exceeds the available supply. Diversion of water has increasingly come into question because once the water is diverted, it is highly unlikely that the receiving area will ever relinquish these augm ented waters. There are a host of problems that emerge as one considers sustainability and water resources. These include water quality and health, ground water management and use, environmental protection of source waters, transport and flood protection, natural disaster problems that include the loss of essential water services when areas experience hurricanes or earthquakes or sea level rise.

The problem of protecting against low probability events with severe consequences must receive additional attention. These elements of sustainability and water resources all have a common element, namely anticipation of change. Changes occur in water systems as the physical infrastructure ages. Changes are occurring in both the demand and supply of water resources. These changes will certainly continue into the future. It is very important that our water systems now and in the future incorporate the ability to adjust and remain viable in the face of unknown future stresses and changes.

There have been more modest increases in the percentage of urban populations now served for water supply and sanitation. While the absolute numbers of rural population not served by water or sanitation have decreased during this decade over the developing regions, the absolute numbers of population in urban areas not served by water supply and sanitation have actually increased. This outcome reflects the population driver as rural people leave the countryside to move to urban areas. The data for Africa is particularly important.

It shows that the absolute numbers of unserved peoples in both urban and rural areas have increased over the past decade. In this region, the population driver has outpaced the provision of water and sanitation services. In the Middle East, the data shows that more progress has been achieved in the urban areas than in the rural areas.

It is important to note that Libya is withdrawing water at over 400% of its annual renewable water resources. This is clearly not sustainable; it is mining water from ground water under the desert and pumping it to the coast for human use needs. Examine the per capita withdrawn (cubic meter/person) in comparison with the overall world average. Egypt, Sudan, and China all use 87% or more of their water withdrawn for agricultural purposes. Here the population driver coupled with the relatively fixed fresh water supply is clearly evident. The data shows a range of 30 cubic meters per capita per year (Egypt) to 9,940 cubic meters per capita per year for the United States. In the case of Egypt, the country is entirely dependent upon the inflow of the Nile River from the Sudan. The countries in the Middle East all have in common a very limited annual internal renewable water resource. Water is the life-blood for these countries and effective shared management of this resource is essential for sustainability.

Increasing salinity of irrigated croplands, results from the high evaporation of irrigation water applied in these areas. The salts are left behind as the irrigation water evaporates. As the salinity increases, the soil becomes less productive and it requires crops with a high tolerance for salinity.

Observations

There are a number of important observations that we may draw from the information presented in this chapter. These observations include the following:

- It is imperative that we seek and implement the most effective and efficient use of water to meet human and environmental needs.
- Additional resources need to be committed to provide water to meet basic human requirements for health and survival.
- There is an important need to address issues of equity in terms of the use of limited water resources.
- It is important to recognize the uncertainty that exists regarding the potential adverse impacts of global climate change on fresh water resources.

- It is imperative to educate and train the water resource professionals who will provide the knowledge, skills, and leadership to meet the multiple water resource challenges of the 21st century.
- It is very important to educate the public as to the value and worth of water in society.
- There is a great need to utilize the watershed approach on an appropriate scale to ensure the most effective planning and management of water resources in the watershed.
- In order to build the necessary trust to implement important water projects, the stakeholders must be involved in meaningful ways in the planning and decision processes.
- It is critical to provide the most effective and innovative institutional arrangements possible to ensure effective implementation and sustained operation of key water resource facilities.

In autumn 2002, thousands of decomposing chinook salmon, a threatened species under the Endangered Species Act, (1) lined the dry river bed of the Klamath River and permeated the air with the smell of unnecessary death. It was the worst fish kill in American history, with 34,000 to 70,000 adult fish carcasses lining the banks of the Klamath River for thirty miles. Although these salmon perished due to two fish pathogens, a contributing factor was a shortage of water in the Klamath River. In the last five years, virtually every region of the United States has experienced a water shortage and, by 2013, at least thirty-six states anticipate some sort of water shortage. Shortages occur because virtually every aspect of American life ties itself to water, with vast amounts devoted to producing electricity, growing food, manufacturing household goods, and serving other personal uses.

Although water shortages have long occurred in the United States, evidence suggests these shortages will continue to worsen, especially in the arid West. Inadequate implementation of the doctrine of prior appropriation, the most popular water system in the western United States, exacerbates this continued decline. First, the doctrine gives priority to the earliest water users; thus, if a shortage occurs, the state dispenses water in the order it granted permits and cuts off the most recent users. Second, a water user may divert and use as much water as the diverter can put to "beneficial use." Thus, under the prior appropriation system, the state grants each user a certain allocation of water which the user may not exceed, and which must be put to beneficial use. However, because states largely have failed to measure the amount of water a user diverts, a process known as "source metering," it is unclear whether users are complying with the conditions of their state permits.

In 1993, Washington became the first western state to require the measurement of virtually all surface water withdrawals, allowing the state to determine the amount of water diverted from rivers and effectively manage water supplies.

The Washington legislature created this source-metering law, part of a larger package aimed at promoting salmon recovery, to ensure compliance with water appropriation permits, protect instream uses, and help determine whether the state has water available for appropriation. After fifteen years and multiple revisions, the statute appears to have made the state's management of water more efficient. However, no other western state has followed in Washington's footsteps. In light of climate change, all western states should adopt source metering. One study suggests that in the West, "no other effect of climate disruption is as significant as how it endangers... already scarce... water supplies." Climate change influences water supplies mainly by affecting precipitation. As temperatures rise, less snow falls in the West and snowpacks shrink, making less water available. In Washington, on the other hand, source metering has helped to leave more than 300,000 acre feet of water in streams. Because source metering leaves more water in streams, it may be the most effective tool to ensure the efficient functioning of the prior appropriation doctrine in a climate-changing world. Kansas and Texas have implemented some form of source metering, mainly to deal with water shortages. Additionally, WaterWatch of Oregon, a fiver conservation group in the western state of Oregon, proposed a state bill to require source metering throughout the state, but the bill died in the 2007 state legislative session. Curiously, the western states, notoriously plagued by water shortages, seem the most resistant to source metering laws, even though they have the most to gain from efficient management of water resources.

Water Consumption and Waste

In the last fifty years, world water consumption has tripled. Water use in the western United States continues to rise as the demand for water to provide energy and support agricultural and metropolitan uses, recreation, fish and wildlife habitat, and water quality protection increases. The climate of the West, with its arid, desert-like lands that receive less than twenty inches of rainfall per year, exacerbates water shortage problems. However, even in the wet areas of western Washington and Oregon, where rainfall can exceed more than 100 inches per year, water shortages have become a concern.

Agricultural use

An estimated 408 billion gallons of water were withdrawn in the United States for all uses in 2000. Of this amount, more fresh water is used in agriculture than for any other use. Irrigation uses the overwhelming majority of water consumed in western states. For example, irrigation withdrawals consume 80% of all water used in Utah and 90% of all the water used in New Mexico. Additionally, many crops grown in the West are low-value crops. In California, pasture, alfalfa, cotton and rice—the four largest water-using crops—use over 50% of all agricultural water. However, the economic value of all these crops together is similar to that of the state's grape crop, which uses only one-ninth as much water. History is one reason for such inefficiency: states granted very generous water rights to early farmers, especially those raising livestock, and the farmers continue to pass down these property rights in water through generations. Existing water users have essentially fully appropriated all available water, meaning that new users can only obtain water when prior existing uses change. Although agriculture traditionally employed more than half of the western population, this number has been on the decline. By 1991, the natural resource industries together provided less than 6% of employment in the West. As agriculture employs fewer people, and the demand for water remains high, water use has shifted away from agricultural use and toward economically higher-valued industrial and municipal uses. Farmers have begun to market their water rights, realizing they can make more money selling their water supply than by growing low-value, water-intensive crops like alfalfa and rice.

Urban Use

Population in the West has exploded over the last few decades. Spatial changes have been the most significant, with people moving away from rural areas and congregating in urban areas. As western metropolitan areas grow, so does the demand for water. In fact, the most recent report from the United States Geological Survey found that municipal withdrawals increased by 8% between 1995 and 2000 alone. Homes account for more than half of the municipal withdrawals, representing much greater consumption than either business or industry. Location also causes these amounts to increase. For example, the arid West has very high per capita residential water use, due to landscape irrigation.

As more water moves toward urban use, the use becomes less elastic because a municipality must always provide water for the basic needs of its citizens. On the other hand, a farmer can forgo applying water to his crops during a year of shortage. Projections suggest that people will continue moving to the West for at least the next twenty-five years, meaning the demand on water will remain high and become less elastics.

Salmon

In Washington and throughout the Pacific Northwest, salmon have played a critical role in history, culture, economy, and recreation. Tribal people value salmon for subsistence and their cultural significance, fishermen value them for sport and economic importance, and environmentalists value them for their ecological significance. Habitat loss, however, poses a considerable threat to the existence of salmon.

Wild salmon declined drastically during the twentieth century; by 1999, salmon had disappeared from 40% of their historic spawning grounds in Washington, Oregon, California, and Idaho. Water shortages often cause habitat loss because salmon need clean, cool water in order to survive. Low streamflows can interfere with upstream migration and may reduce or even eliminate spawning habitats. In 1991, the federal government placed one population of Washington salmon on the list of Endangered Species List, partially due to lack of habitat. Over the next eight years, the federal government listed three additional populations of salmon in Washington as endangered or threatened. Lack of adequate salmon habitat played a dominant role in spurting the government to create these listings. Recognizing that salmon populations would not thrive without drastic changes, the Washington legislature enacted a legislative package to address salmon recovery. One of the new laws established a requirement that all new and certain existing surface water users within the state measure their surface water diversions. In response to these requirements, Washington became the first western state to implement source-metering at the statewide level.

The slow implementation of Washington's 1993 source metering statute prompted a lawsuit in 1999 to spur completion. As a result, the Washington State Department of Ecology (WDOE) was ordered to submit a compliance plan. Thus, fifteen years after its enactment and several lawsuits later, the statute finally seems to have achieved its goal of efficient state water management.

The Beginning of the Source Metering Statute

The 1993 law required all new and certain existing water users to measure surface water diversions as part of a larger salmon recovery package. The law required measurement of every 1) new surface water permit, 2) existing surface water permit exceeding one cubic foot per second (cfs), and 3) all new and existing permits, regardless of size, in areas where salmon stocks are "depressed or in critical condition." Essentially, the only water users not regulated by the new statute consisted of those existing surface water users with permits of less than one cfs outside of critical salmon areas. The legislature intended this statute to be the first step toward effective water management. According to the legislature, water management requires information gathering, meaning the state must know who is using what amount of water and when they are using it. Without this information, the WDOE could not efficiently identify any illegal uses of water, nor could it provide adequate water for fish, resolve conflicts between water uses, or promote conservation.

Although state law required source metering in 1993, the WDOE failed to adopt implementing regulations. Without new regulations, the existing regulations did not require, or even allow, metering of any kind. The WDOE also failed to apply the source metering rule to groundwater. Although the original statute applied only to surface water, the statutory provisions regulating groundwater incorporated all surface water regulations into the groundwater code. Thus, when the legislature required source metering for surface water, the rules should have applied to groundwater as well.

In 1999, as a response to these deficiencies, a coalition of environmental groups filed suit, contending that the WDOE failed to implement administrative rules regarding water diversions, particularly implementing the requirement to establish water metering. The Thurston County Superior Court agreed, ordering the WDOE to create a compliance plan that included adopting either a new or revised administrative rule requiring source metering by December 31, 2001. In addition, the court ordered the WDOE to require the metering of 80% of water use in each of the sixteen critical fish basins by December 31, 2002.

The WDOE responded by creating a compliance plan and adopting a revised administrative rule on source metering by the required deadline. By 2003, the WDOE issued administrative orders requiring metering for 903 water rights, which amounted to 80% of the total estimated water diversions in the critical fish basins. In order to ensure that water users met these requirements, the state legislature appropriated over $3 million to defray the installation costs of the metering equipment. Moreover, the WDOE's rules required measuring devices on all new water rights for both surface and groundwater withdrawals and source metering for all water users requesting a change or extension to an old right.

2

Irrigation Management

Irrigation is the artificial *exploitation* and *distribution* of water at *project level* aiming at *application* of water at *field level* to agricultural crops in dry areas or in periods of scarce rainfall to assure or improve crop production.

This article is about organizational forms and means of management of irrigation water at project level.

Water Management

The most important physical elements of an *irrigation project* are *land* and *water*. In accordance with the propriety relations of these elements there may be different types of water management :

- the communal type
- the enterprise type
- the utility type.

Communal Type

Until the end of the 19th century the development of irrigation projects occurred at a mild pace, reaching a total area of some 50 million ha worldwide, which is about 1/5 of the present area. The land was often private property or assigned by the village authorities to male or female farmers, but the water resources were in the hands of clans or communities who managed the water resources *cooperatively*.

Enterprise Type

The enterprise type of water management occurred under large landowners or agricultural corporations, but also in centrally controlled societies. Both the land and water resources are in one hand.

Large plantations were found in colonised countries in Asia, Africa, and Latin America, but also in countries employing slave labour. It concerned mostly the large scale cultivation of commercial crops such as bananas, sugarcane and cotton. As a result of land reforms, in many countries the estates were reformed into a cooperatives in which the previous employers became members and exercised a cooperative form of land and water management.

Utility Type

Irrigation canals of the Gezira Scheme, Sudan, from space, 1997, with the utility type of management. The water comes from the Blue Nile.

The utility type of water management occurs in areas where the land is owned by many, but the exploitation and distribution of the water resources are managed by (government) organizations.

After 1900 governments assumed more influence over irrigation because :

- water was increasingly considered government property owing to the increasing demand for good quality water and the reducing availability
- governments embarked on large scale irrigation projects as they were considered more efficient
- the development of new irrigation schemes became technically, financially and organizationally so complicated that they fell outside the capabilities of the smaller communities
- the import and export policies of governments required the cultivation of commercial cash crops whilst, by controlling the water management, the farmers could be more easily guided to plant these kind of crops.

The water management signified a large subsidy on irrigation schemes. From 1980 the operation and maintenance of many irrigation projects was gradually handed over to water user organizations (WUA's) who were to assume these tasks and a large part of the costs, whereby the water rights of the members had to be respected.

The exploitation of water resources via large storage dams-that often provided electric power as well-and diversion weirs normally remained the responsibility of the government, mainly because environmental protection and safety issues were at stake. In the past, the utility type of water management witnessed more conflicts and disturbances then the other types.

Tariffs

Irrigation water has a price by which the management costs must be covered. The following tariff (water charge) systems exist :

- No tariff, the government assumes the costs
- Tariff in labour hours, which holds mainly in communal types of management in traditional irrigation systems
- Yearly area tariff, a fixed price per ha per year
- Seasonal area tariff, a fixed price per ha per season with the higher price in the dry season
- Volumetric tariff, a fixed price per m3 of water; the consumption is measured by water meters
- Block or stepped-up pricing for water use per ha; the price increases as the water consumption per ha falls in a higher block.

The use of groundwater for irrigation is often licensed by government and the well owner may be permitted to withdraw only a maximum volume of water per year at a certain price.

Cost Recovery

The recovery of water charges may be below target, because :

- The revenues accrue to a (government) organization other than the one responsible for the management
- Farmers and water users have no say in the water management
- Lack of communication between farmers and project managers
- Poor farmers are unable to comply
- Farmers do not receive water according to need; for example insufficient quantity and/or inappropriate time
- Corruption at management level.

Cost Coverage

The cost recovery is often insufficient for full cost coverage, for example:

Country	*Cost recovery (%)*	*Cost coverage (%)*	*Remarks*
Argentine	67	12	low tariff: $70/ha/year
Bangladesh	3-10	<1	tariffs not enforced
Brazil, Jaiba project	66	52	
Colombia	76	52	
Turkey	76	30-40	
Sri Lanka	8	<1	tariffs not enforced

WUA's

From 1980 programs were developed to transfer the operation and maintenance tasks from the government to water user associations (WUA's) that show some resemblance to water boards in the Netherlands, with the difference that it concerns irrigation rather than drainage and flood control. An effective development occurred in Mexico, where in 1990 a program of WUA's was initiated with tradable water rights. By 1998 some 400 WUA's were in operation commanding on average 7600 ha per WUA. They were able to recover more than 90% of the tariffs, mainly because they had to be paid in advance. Government subsidies to the water distribution and maintenance reduced to only 6%.

Water Delivery Principles

In large irrigation schemes, the distribution of irrigation water and the delivery at the farm gate is often arranged by *rotational turns* (e.g. every fortnight). The quantity of water to be received is often proportional to the farm size. As the canals usually transport constant flows, the water is being received during a period of time proportional to the farm size (e.g. every fortnight during 2 hours).

In smaller irrigation schemes the water delivery may be arranged "on demand" with water charges are on a volumetric basis. This requires a precise bookkeeping system. As the demand may be fluctuating over time, the distribution system and infrastructure is relatively expensive because it must be able to cope with periods of peak demand. During periods of water scarcity, negotiations are due to regulate the supply. From point of view of efficient irrigation water-use this is the most effective system. In projects with an uncertain supply of water due to annual variations in river discharge, water users at the top (the head users) of the irrigation system (i.e. near the system's take-off point) often have preference, to a certain extent, over users at the tail-end. Hence, the number of farmers that are able to grow an irrigated crop may vary from year to year according to the riparian water rights.

In regions with a structural water scarcity, the principle of *water duty* is often applied, whereby the duty per ha per season is only a fraction of the full irrigation need per ha (i.e. the *irrigation intensity* is less than 100%). Thus, farmers can irrigate only part of their land or irrigate their crops with a limited amount of water, whereby they may choose between crops with a high consumptive use (e.g. rice, sugarcane, most orchards) or a low consumptive use (e.g. cereals, cotton). In India, such practice is called *protective irrigation*, which aims at equal distribution of scarce means and prevention of acute famine. Owing to competition for water, the water delivery practices may deviate from the principles.

Water Delivery Practices

In practice the distribution of irrigation water is subject to competition. Influential farmers may be able to acquire more water than they are entitled to. Water users at the upstream part of the irrigation system can more easily intercept extra water than the tail-ender. The degree of farmers' influence is correlated to the relative position of their land in the topography of the scheme.

Tail-end Problems

R.Chambers cites authors who have reported tail-end problems.

Examples are:

- The old Sardar canal project in the state of Gujarat, India, was designed with an irrigation intensity of 32%, but at the upstream part the delivery was at an intensity of 42% (i.e. 131% of the design norm) and at the downstream end it was only 19% (i.e. 59% of the norm), although the project aimed at protective irrigation with equal rights for all.
- The Sardar Sahayak Pariyojana irrigation project, an extension of the Sardar canal project with 1.7 million ha, the head farmers received 5 times more water than the tail-enders, although the project was designed for equal distribution of the scarce water.
- The Ghatampur distributary canal in the Ramganga irrigation project in the state of Uttar Pradesh, India, delivered an amount of water equal to 155% of the design discharge to the Kisarwal district canal near the head of the distributary and only 22% to the Bairampur district canal at the downstream end.

Environmental Impact of Irrigation

Environmental impacts of irrigation are the changes in quantity and quality of soil and water as a result of irrigation and the ensuing effects on natural and social conditions at the tail-end and downstream of the irrigation scheme.

The impacts stem from the changed hydrological conditions owing to the installation and operation of the scheme.

An irrigation scheme often draws water from the river and distributes it over the irrigated area. As a hydrological result it is found that:

- the downstream river discharge is reduced
- the evaporation in the scheme is increased
- the goundwater recharge in the scheme is increased
- the level of the water table rises
- the drainage flow is increased.

These may be Called Direct Effects

The effects thereof on soil and water quality are indirect and complex, Water logging and soil salination are part of these, whereas the subsequent impacts on natural, ecological and socio-enonomic conditions is very intricate.

Irrigation can also be done extracting groundwater by (tube) wells. As a hydrological result it is found that the level of the water descends. The effects may be water mining, land/soil subsidence, and, along the coast, saltwater intrusion.

Irrigation projects can have large benefits, but the negative side effects are often overlooked.

Reduced Downstream River Discharge

The reduced downstream river discharge may cause:

- reduced downstream flooding
- disappearance of ecologically and economically important wetlands or flood forests
- reduced availability of industrial, municipal, household, and drinking water
- reduced shipping routes. Water withdrawal poses a serious threat to the Ganges. In India, barrages control all of the tributaries to the Ganges and divert roughly 60 percent of river flow to irrigation
- reduced fishing opportunities. The Indus River in Pakistan faces scarcity due to over-extraction of water for agriculture. The Indus is inhabited by 25 amphibian species and 147 fish species of which 22 are found nowhere else in the world. It harbours the endangered Indus River dolphin, one of the world's rarest mammals. Fish populations, the main source of protein

and overall life support systems for many communities, are also being threatened

- reduced discharge into the sea, which may have various consequences like coastal erosion (e.g. in Ghana) and salt water intrusion in delta's and estuaries. Current water withdrawal from the river Nile for irrigation is so high that, despite its size, in dry periods the river does not reach the sea. The Aral sea has suffered an "environmental catastrophe" due to the interception of river water for irrigation purposes.

Increased Groundwater Recharge, Waterlogging, Soil Salinity

This illustrates an environmental impact of upstream irrigation developments causing an increased flow of groundwater to this lower lying area leading to the adverse conditions. The increased groundwater recharge stems from the unavoidable deep percolation losses occurring in the irrigation scheme. The lower the irrigation efficiency, the higher the losses. Although fairly high irrigation efficiencies of 70% or more (i.e. losses of 30% or less) can be obtained with sophisticated techniques like sprinkler irrigation and drip irrigation, or by precision land levelling for surface irrigation, in practice the losses are commonly in the order of 40 to 60%. This may cause:

- rising water tables,
- increased storage of groundwater that may be used for irrigation, municipal, household and drinking water by pumping from wells,
- waterlogging and drainage problems in villages, agricultural lands, and along roads with mostly negative consequences. The increased level of the water table can lead to reduced agricultural production.
- shallow water tables are a sign that the aquifer is unable to cope with the groundwater recharge stemming from the deep percolation losses,
- where water tables are shallow, the irrigation applications are reduced. As a result, the soil is no longer leached and soil salinity problems develop,
- stagnant water tables at the soil surface are known to increase the incidence of water borne diseases like malaria, filariasis, yellow fever, dengue, and schistosomiasis (Bilharzia) in many areas. Health costs, appraisals of health impacts and mitigation measures are rarely part of irrigation projects, if at all.

- to mitigate the adverse effects of shallow water tables and soil salinization, some form of watertable control, soil salinity control, drainage and drainage system is needed.
- As drainage water moves through the soil profile it may dissolve nutrients (either fertilizer based or naturally occurring) such as nitrates, leading to a built up of those nutrients in the ground water aquifer. High nitrate levels in drinking water can be harmful to humans particularly for infants under 6 months where it is linked to 'blue-baby syndrome'.

Case Studies:

1. In India 2.189.400 ha have been reported to suffer from waterlogging in irrigation canal commands. Also 3.469.100 ha were reported to be seriously salt affected here
2. In the Indus Plains in Pakistan, more than 2 million hectares of land is waterlogged. The soil of 13.6 million hectares within the Gross Command Area was surveyed, which revealed that 3.1 million hectares (23%) was saline. 23% of this was in Sindh and 13% in the Punjab. More than 3 million ha of water-logged lands have been provided with tube-wells and drains at the cost of billions of rupees, but the reclamation objectives were only partially achieved. The Asian Development Bank (ADB) states that 38% of the irrigated area is now waterlogged and 14% of the surface is too saline for use
3. In the Nile delta of Egypt, drainage is being installed in millions of hectares to combat the water-logging resulting from the introduction of massive perennial irrigation after completion of the High Dam at Assuan
4. In Mexico, 15% of the 3.000.000 ha if irrigable land is salinized and 10% is waterlogged
5. In Peru some 300.000 ha of the 1.050.000 ha of irrigable land suffers from this problem.
6. Estimates indicate that roughly one-third of the irrigated land in the major irrigation countries is already badly affected by salinity or is expected to become so in the near future. Present estimates for Israel are 13% of the irrigated land,, Australia 20%, China 15%, Irak 50%, Egypt 30%. Irrigation-induced salinity occurs in large and small irrigation systems alike
7. FAO has estimated that by 1990 about 52 x 10^6 ha of irrigated land will need to have improved drainage systems installed, much of it subsurface drainage to control salinity.

Reduced Downstream Drainage and Groundwater Quality:

- The downstream drainage water quality may deteriorate owing to leaching of salts, nutrients, herbicides and pesticides. This may negatively affect the health of the population at the tail-end and downstream of the irrigation scheme, as well as the ecological balance. The Aral sea, for example, is seriously polluted by drainage water.
- The downstream quality of the groundwater may deteriorate in a similar way as the downstream drainage water and have similar consequences.

Reduced Downstream River Water Quality

Owing to drainage of surface and groundwater in the project area, which waters may be salinized and polluted by agricultural chemicals like biocides and fertilizers, the quality of the river water below the project area can deteriorate, which makes it less fit for industrial, municipal and household use. It may lead to reduced public health. Polluted river water entering the sea may adversely affect the ecology along the sea shore.

Affected Downstream Water Users

Downstream water users often have no legal water rights and may fall victim of the development of irrigation.

Pastoralists and nomadic tribes may find their land and water resources blocked by new irrigation developments without having a legal recourse.

Flood-recession cropping may be seriously affected by the upstream interception of river water for irrigation purposes.

- In Baluchistan, Pakistan, the development of new small-scale irrigation projects depleted the water resources of nomadic tribes travelling annually between Baluchistan and Gujarat or Rajasthan, India
- After the closure of the Kainji dam, Nigeria, 50 to 70 per cent of the downstream area of flood-recession cropping was lost.

Lost Land use Opportunities

Irrigation projects may reduce the fishing opportunities of the original population and the grazing opportunities for cattle. The livestock pressure on the remaining lands may increase considerably, because the ousted traditional pastoralist tribes will have to find their subsistence and existence elsewhere, overgrazing may increase, followed by serious erosion and the loss of natural resources. The Manatali reservoir formed by the Manantali dam in Mali intersects the migration routes of nomadic pastoralists and destroyed 43000 ha of savannah, probably leading to overgrazing and erosion elsewhere. Further, the reservoir destroyed 120 km^2 of forest. The depletion of groundwater aquifers, which is caused by the suppression of the seasonal flood cycle, is damaging the forests downstream of the dam.

Groundwater Mining with Wells, Land Subsidence

When more groundwater is pumped from wells than replenished, storage of water in the aquifer is being mined. Irrigation from groundwater is no longer sustainable then. The result can be abandoning of irrigated agriculture.

The hundreds of tubewells installed in the state of Uttar Pradesh, India, with World Bank funding have operating periods of 1.4 to 4.7 hours/day, whereas the were designed to operate 16 hours/day.

In Baluchistan, Pakistan, the development of tubewell irrigation projects was at the expense of the traditional qanat or karez users.

Groundwater-related subsidence of the land due to mining occurred in the USA at a rate of 1m for each 13m that the watertable was lowered.

Homes at Greens Bayou near Houston, Texas, where 5 to 7 feet of subsidence has occurred, were flooded during a storm in June 1989 as shown in the picture.

Irrigation in Viticulture

The role of irrigation in viticulture is considered both controversial and essential to wine production. In the physiology of the grapevine, water is a vital component to function of the vine with its presence or lack impacting photosynthesis, new plant shoot growth, as well as the development of grape berries. While climate and humidity play important roles, the typical grape vine needs an average of 27 inches (680 millimetres) of water a year, ideally occurring during the winter and spring months of the growing season. In many Old World wine regions, natural rainfall is considered the only source for water that will still allow the vineyard to maintain its *terroir* characteristics. The practice of irrigation is viewed by some critics as unduly manipulative with the potential for detrimental wine quality due to high yields that can be artificially increased with irrigation. It has been historically banned by the European Union's wine laws, though in recent years individual countries (such as Spain) have been loosening their regulations and France's wine governing body, the *Institut National des Appellations d'Origine* (INAO), has also been reviewing the issue.

In very dry climates that receive little rainfall, irrigation is considered essential to any viticultural prospects. Many New World wine regions such as Australia and California regularly practice irrigation in areas that couldn't otherwise support viticulture. Advances and research in these wine regions (as well as some Old World wine regions such as Israel), have shown that potential wine quality could increase in areas where irrigation is kept to a minimum and managed. The main principle behind this is controlled water stress where the vine receives sufficient water during the budding and flowering period that is then scaled back during the ripening period where the vine then responds by funneling more its limited resources into developing the grape clusters instead of excess foliage. If the vine receives too much water stress, then photosynthesis and other important process could be impacted with vine essentially shutting down. The availability of irrigation means that if drought conditions emerge, sufficient water can be provided for the plant so that the balance between water stress and development is kept to optimal levels.

History

The practice of irrigation has had a long history in wine production. Archaeologists describe it as one of the oldest practices in viticulture, with irrigation canals discovered near vineyard sites in Armenia and Egypt dating back more than 2600 years. Irrigation was already widely practice for other agricultural crops since around 5000 BC. It is possible that the knowledge of irrigation helped viticulture spread from these areas to other regions due to the potential for the grapevine to grow in soils too infertile to support other food crops. A somewhat hardy plant, the grapevine largest need is for sufficient sunshine and is able to flourish with minimum needs of water and nutrients. In areas where its water needs are unfulfilled, the availability of irrigation meant that viticulture could still be supported.

In the 20th century, the growing wine industries of California, Australia and Israel were greatly enhanced by advances in irrigation. With the development of more cost efficient and less labour intensive ways of watering the vines, vast tracks of very sunny but dry lands were able to be converted into growing wine regions. The ability to control the precise amount of water each vine received allowed producers in these New World wine regions to develop styles of wines that could be fairly consistent each year regardless of normal vintage variation. This created a stark contrast to the Old World wine regions of Europe where vintage variation, including rainfall, had a pronounced effect on the potential wine style each year. Continuing research explored the way that controlled (or supplemental) irrigation could be used to increase potential wine quality by influence how the grapevine responds to its environment and funnels resources into developing the sugars, acids and phenolic compounds that contribute to a wine's quality. This research lead to the development of ways to measure the amount water retention in the soil to where individual irrigation regimes could be plotted for each vineyard that maximized the benefits of water management.

Role of Water in Viticulture

The presence of water is essential for the survival of all plant life. In a grapevine, water acts as a universal solvent for many of the nutrients and minerals needed to carry out important physiological functions-which the vine receives by absorbing the nutrient-containing water from the soil. In the absence of water in the soil, the root system of the vine may have difficulties absorbing these nutrients. Within the structure of the plant itself, water acts as a transport within the xylem bring these nutrients to all ends of the plant. During the process of photosynthesis, water molecules combine with carbon derived from carbon dioxide to form glucose which is the primary energy source of the vine as well as oxygen as a by-product.

In addition to its use in photosynthesis, a vine's water supply is also depleted by the processes of evaporation and transpiration. In evaporation, heat (aided by wind and sunlight) causes water moisture in the soil to evaporate and escape as vapor molecules. This process is inversely related to humidity with evaporation often taking place at faster rates in areas with low relative humidity. In transpiration, this evaporation of water occurs directly in the wine as water is released from the plant through the stomata that is located underneath the leafs of a grape vine. This process helps the vine combat against the effects of heat stress which can severely damage the physiological functions of the vine (somewhat similar to how perspiration works with humans and animals). The presence of adequate water in the vines can help keep the internal temperature of a the leaf only a few degrees above the temperature of the surrounding air. However, if water is severely lacking then that internal temperature could jump nearly 18 °F (10 °C) warmer than the surrounding air which leads the vine to develop heat stress. The duel effects of evaporation and transpiration is called evapotranspiration. A typical vineyard in a hot, dry climate can lose as much as 40,000 litres (11,000 U.S. gal; 8,800 imp gal) of water through evapotranspiration during the growing season. Even in a cooler, dry climate, a vineyard could still lose enough water to flood the vineyard with 16 inches (400 mm) if all that water was suddenly replaced at once.

Factors Influencing Irrigation

Climates with low humidity promote faster rates evapotranspiration which reduce the grapevine's water supply. These areas may need to utilize supplemental irrigation.

There are essentially two main types irrigation-primary irrigation, which is needed for areas (such as very dry climates) that lack sufficient rainfall for viticulture to even exist, and supplemental irrigation where irrigation is used to "fill in the gaps" of natural rainfall to bring water levels to more optimal numbers as well as to serve as a preventive measure in case of seasonal drought conditions. In both cases, both the climate and the vineyard soils of the region will play an instrumental role in irrigation's use and effectiveness.

Impact of Different Climate Types

Viticulture is most commonly found in Mediterranean, continental and maritime climates with each unique climate providing its own challenges in providing sufficient water at critical times during the growing season. In Mediterranean climates irrigation is usually needed during the very dry periods of the summer ripening stages where drought can be a persistent threat. The level of humidity in a particular macroclimate will dictate exactly how much irrigation is needed with high levels of evapotranspiration more commonly occurring in Mediterranean climates that have low levels of humidity such as part of Chile and the Cape Province of South Africa. In these low humidity regions, primary irrigation maybe needed but in many Mediterranean climates the irrigation is usually supplemental.

Continental climates are usually seen in areas further inland from the coastal influences of oceans and large bodies of water. The difference from the average mean temperature of its coldest and hottest months can be quite significant with moderate precipitation that usually occurs in the winter and early spring. Depending on the water retaining ability of the soil the grapevine may receive enough water during this period to last throughout the growing season with little if any irrigation needed. For soils with poor water retention, the dry summer months may require some supplemental irrigation. Examples of continental climates that use supplemental irrigation include the Columbia Valley of Washington State and the Mendoza wine region of Argentina.

Maritime climates tend to fall between Mediterranean and continental climates with a moderate climate that is tempered by the effects of a large body of water nearby. As with Mediterranean climates, the humidity of the particular macroclimate will play a significant role in determining how much irrigation is needed. In most cases irrigation, if it is used at all, will only be supplemental in years where drought may be an issue. Many maritime regions, such as Rias Baixas in Galicia, Bordeaux and the Willamette Valley in Oregon, suffer from the diametric problem of having too much rain during the growing season.

Impact of Different Soil Types

Soil can have a significant impact on the potential quality of wine. While geologist and viticulturist are not exactly sure what type of immutable or *terroir* based qualities that soil can impart on wine, there is near universal agreement that a soil's water retention and drainage abilities play a primary role. Water retention refers to the soil's ability to hold water. The term "field capacity" is used to describe the maximum amount of water that deeply moisten soil will retain after normal drainage. Drainage is the ability of water to move freely throughout the soil. The ideal circumstance is soil that can retain sufficient amount of water for the grapevine but drains well enough to where the soil doesn't become water-logged. Soil that doesn't retain water well encourages the vine to easily sleep into water stress while soil that doesn't drain well runs of the risk of water-logged roots being attacked by microbial agents that consume all the soil nutrients and end up starving the vine.

The depth, texture and composition of soils can influence its water retaining and draining ability. Soils containing large amounts of organic material tend to have the highest water retention abilities. These types of soils include deep loams, silty soils like what is typically found on the fertile valley floors such as in the California's Napa Valley. Clay particles have the potential to remain in colloidal suspension for long periods of time when it is dissolved in water. This gives clay-based soils the potential to retain significant amount of water such as the clay soils of the Right bank Bordeaux region of

Pomerol. Many regions with these types of water retaining soils have little need for irrigation, or if they do it is usually supplemental during periods of drought. Soils with poor water retention include sand and alluvial gravel based soils such as those found in the Barolo and Barbaresco zones of Italy or in many areas of South Australia. Depending on the climate and amount of natural rainfall, areas with poor water retention may need irrigation.

Just as having too little water is detrimental to the grapevine, so too is having too much. When vines become water-logged they become a target for various microbial agents such as bacteria and fungi that compete with the vine for nutrients in the soil. Additionally excessively moist soil is poor conductor of valuable heat radiating from the ground. In general wet soils are cold soils which can be especially problematic during the flowering causing poor berry set that could lead to coulure. It also becomes an issue during the ripening stage when vines in cool-climate regions may need additional heat radiated from the ground in order to sufficiently ripen its fruit (an example of this is the slate-based vineyards of the Mosel in Germany). Therefore, well draining soils are considered very conducive to producing quality wine. In general light-textured (such as sand and gravel) and stony soils tend to drain well. Soils heavy soils and those with high proportions of organic matter also have the potential to drain well if they having a crumbling texture and structure. This texture relates to the friability of the soil which can come from earthworms and other organisms that have burrowed tunnels throughout the soil. Much like rocks, these tunnels give water a freer passageway through soil and contributes to its drainage.

Measuring Soil Moisture

Tensiometers can be used to measure soil moisture. The components of this example include (1) porous cup, (2) water-filled tube (3) sensor-head and a (4) pressure sensor.

Because of the problems associated with water-logged and wet soils, it is important for viticulturist to know how much water is currently in the soil before deciding if and how much to irrigate. There are several methods of evaluating soil moisture. The most basic is simple observation and feeling of the soil, however this has its limitations since the subsoil may be moist while the surface soil appears dry. More specific measurements can be attained by using tensiometers which evaluates surface tension of water extracted from the soil. The presence of water in the soil can be measured by neutron moisture meters that utilize an aluminum tube with an internal neutron source that detect the subtle change between the water in the soil. Similarly, gypsum block placed throughout the vineyard contain an electrode that can be used to detect the electrical resistance that occurs as the soil dries and water is released by evaporation. Since the 1990s there has been greater research into tools utilizing time-domain reflectivity and capacitance probes. In addition to monitoring for excessive moisture, viticulturists also keep an eye for signs of water stress due to severe lack of water.

Irrigation Systems

There are several methods of irrigation that can be used in viticulture depending on the amount of control and water management desired. Historically, surface irrigation was the most common means using the gravity of a slope to release a flood of water across the vineyard. In the early history of the Chilean wine industry, flood irrigation was widely practiced in the vineyards using melted snow from the Andes Mountains channelled down to the valleys below.

This method provided very little control and often had the adverse effect of over-watering the vine. An adoption of method was the furrow irrigation system used in Argentina where small channels ran through the vineyard providing irrigation. This provide slightly more control since the initial amount of water entering the channels could be regulated, however the amount that each vine received was sporadic.

Sprinkler irrigation involves the installation of a series of sprinkler units throughout the vineyard, often spaced as several rows about 65 feet (20 m) apart. The sprinklers can be set on a electronic timer and release predetermined amount of water for a set period of time. While this provides more control and uses less water than flood irrigation, like furrow irrigation the amount that each individual wine receives can be sporadic.

The irrigation system that provides the most control over water management, though conversely the most expensive to install, is drip irrigation. This system involved long plastic water supply lines that run down each row of vines in the vineyard with each individual grape vine having its own individual dripper. With this system, a viticulturist can control the precise amount of water that each grapevine gets down to the drop. An adaption of this system, potentially useful in areas where irrigation may be banned, is underground sub-irrigation where precise measurements of water is delivered directly to the root system.

When and how much? Water is very crucial during the early budding and flowering stages but after fruit set *(pictured)*, the amount of water given to the vine may be scaled back in order to promote water stress.

With abundant water, a grapevine will produce shallow root systems and vigorous growths of new plant shoots. This can contribute to a large, leafy canopy and high yields of large grape berry clusters that may not be sufficiently or physiologically ripe. With insufficient water, many of the vine's important physiological structures, including photosynthesis that contributes to the development of sugars and phenolic compounds in the grape, can shut down. The key to irrigation is to provide just enough water for the plant to continuing function without encouraging vigorous growth of new shoots and shallow roots. The exact amount of water will depend on a variety of factors including how much natural rainfall can be expected as well as the water retaining and drainage properties of the soil.

Water is very crucial during the early budding and flowering stages of the growing season. In areas where there is not sufficient rainfall, irrigation may be needed during this time in the spring. After fruit set, the water needs for the vine drop and irrigation is often withheld till the period of *veraison* when the grapes begin to change colour. This period of "water stress" encourages the vine to concentrate its limited resources into lower yields of smaller berries creating a favourable skin to juice ratio that is often desirable in quality wine production. The benefits or disadvantageous of irrigation during the ripening period itself is a matter of debate and continuing research in the wine growing community. The only area of mostly agreement is the disadvantages of water close to harvest after a prolong dry period. Grapevines that has been subjugated to prolong water stress have a tendency to rapidly absorb large amounts of water if its provided. This will dramatically swell the berries, potentially causing to them crack or burst which will make the prone to various grape diseases. Even if the berries do not crack or burst, the rapid swelling of water will cause a reduce concentration in sugars and phenolic compounds in the grape producing wines with diluted flavours and aromas.

Water Stress

One of the goals of controlled, mild water stress is to discourage the formation of excess new plant growths *(a bud pictured)* which will compete with the developing grape clusters for the vine's limited resources. The term water stress describes the physiological states that grapevines experience when they are deprived of water. When a grapevine goes into water stress one of its first functions is to reduce the growth of new plant shoots which compete with the grape clusters for nutrients and resources. The lack of water also keeps the individual grape berries down to a smaller size which increase its skin to juice ratio. As the skin is filled with color phenolase, tannin and aroma compounds, the increase skin to juice ratio is desirable for the potential added complexity the wine may have. While there is disagreement over exactly how much water stress if beneficial in development grapes for quality wine production, most viticulturist agree that some water stress can be beneficial. The grapevines in many Mediterranean climates such as Tuscany in Italy and the Rhone Valley in France experience natural water stress due to the reduce rainfall that occurs during the summer growing season.

At the far extreme is severe water stress which can have detrimental effects on both the vine and on potential wine quality. To conserve water, a vine will try to conserve water by limiting its lose through transpiration. The plant hormone abscisic acid triggers the stomata on the underside of the plant leaf to stay close in order to reduce the amount of water that is evaporated. While conserving water this also has the consequences of limiting the intake of carbon dioxide needed to sustain photosynthesis.

If the vine is continually stressed it will keeps it stomata closed for longer and longer periods of time which can eventually cause photosynthesis to stop all together. When a vine has been so deprived of water it can exceed what is known as its permanent wilting point. At this point, the vine can become permanently damaged beyond recovery even if later watered. Viticulturists will carefully watch the plant for signs of severe water stress. Some of the symptoms include:

- Flaccid and wilting tendrils
- (During Flowering) Flower clusters that are dried out
- Wilting of young grape leaves followed by maturer leaves
- Chlorosis signalling that photosynthesis has stopped
- Necrosis of dying leaf tissue which leads to premature leaf fall
- Finally, the grape berries themselves start to shrivel and fall off the vine.

The effectiveness of water stress is an area of continuing research in viticulture. Of particular focus is the connection between yield size and the potential benefits of water stress. Since the act of stressing the vine does contribute to reduce photosynthesis-and by extension, reduce ripening since the sugars produced by photosynthesis is needed for grape development-it is possible that a stressed vine with high yields will only produce lots of under ripe grapes. Another interest of study is the potential impact on white grape varieties with enologists and viticulturists such as Cornelius Van Leeuwen and Catherine Peyrot Des Gachons contending that white grape varieties lose some of their aromatic qualities when subjugated to even mild forms of water stress.

Partial Rootzone Drying

In partial rootzone drying, half the roots are allowed to dehydrate which sends signals to the vine that is experiencing "water stress". Meanwhile the irrigated roots on the other side of vine continue to provide sufficient amounts of water so that vital functions like photosynthesis does not cease.

One irrigation technique known as partial rootzone drying (or PRD) involves "tricking" the grapevine into thinking it is undergoing water stress when it is actually receiving sufficient water supply. This is accomplished by alternating drip irrigation to where only one side of the grapevine receives water. The roots on the dry side of the vine produce abscisic acid that triggers some of the vine's physiological responses to water stress-reduced shoot growth, smaller berries size, etc. But because the vine is still receiving water on the other side the stress doesn't become so severe to where vital functions such as photosynthesis is compromised.

Criticism and Environmental Issues

The practice of irrigation has it share of criticism and environmental concerns. In many European wine regions the practice is banned under the belief that irrigation can be detrimental to quality wine production. However, in the early 21st century some European countries have relaxed their irrigation laws or re-evaluated the issue. Of the criticisms levelled towards irrigation, the most common is that it disrupts the natural expression of *terroir* in the land as well as the unique characteristics that comes with vintage variation. In regions that do not practice irrigation, the quality and styles of wines can be dramatically different from vintage to vintage depending on weather conditions and rainfall. Irrigation's contribution to the broader globalization of wine is criticized as promoting a homogenization or "standardization" of wine.

Other criticisms centre around the broader environmental impact of irrigation on both the ecosystem around the vineyard as well as the added strain on global water resources. While advances in drip irrigation has reduced the amount of waste water produced by irrigation, the irrigation of substantial tracts of land in areas like the San Joaquin Valley in California and the Murray-Darling Basin of Australia requires massive amounts of water from dwindling supplies. In Australia, the centuries old practice of flood irrigation used in places like the Murrumbidgee Irrigation Area caused severe environmental damages from water-logging, increase salination and raising the water tables. In 2000, the Australian government invested over 3.6 million AUD into research on how to minimize the damage caused by extensive irrigation. In 2007, concerns about ecological damage to the Russian River caused government officials in California to take similar measures to cut back water supplies and promote more efficient irrigation practices.

Other uses for Irrigation Systems

In addition to providing water for plant growth and development, irrigation systems can also be used for alternative purposes. One of the most common is the dual application of fertilizer with water in a process known as fertigation. Commonly used in drip irrigation systems, this method allows similarly regulate control over how precisely how much fertilizer and nutrients that each vine receives. Another alternative use for sprinkler irrigation systems can occur during the threat of winter or spring time frost. When temperature drop below 32 °F (0 °C), the vine is at risk of developing frost damage that could not only ruining the upcoming years harvest but also kill the vine. One preventive measure against frost damage is to use the sprinkler irrigation system to coat the vines with a protective layer of water that freezes into ice. This layer of ice serves as insulation keeping the internal temperature of the vine from dropping below the freezing mark.

Tidal Irrigation

Tidal irrigation is the subsurface irrigation of levee soils in coastal plains with river water under tidal influence. It is applied in (semi) arid zones at the mouth of a large river estuary or delta where a considerable tidal range (some 2 m) is present. The river discharge must be large enough to guarantee a sufficient flow of fresh water into the sea so that no salt water intrusion occurs in the river mouth.

The irrigation is effectuated by digging *tidal canals* from the river shore into the main land that will guide the river water inland at high tide. For the irrigation to be effective the soil must have a high infiltration capacity to permit the entry of sufficient water in the soil to cover the evapotranspiration demand of the crop. At low tide, the canals and the soil drain out again, which promotes the aeration of the soil.

Irrigation of Alluvial Fans

Irrigation of alluvial fans is the use of water resources, mainly river floods and groundwater recharged by infiltration of river water, to enhance the production of agricultural crops.

- Alluvial fans, when large and flat, are also called *inland delta's*.

General Description

Alluvial fans, also called inland deltas, occur at the foot of mountain ranges and mark the presence of river floods. The rivers flowing at high speed in the mountains carry sediments. Upon losing its speed in the flat land at the foot of the mountain, the water deposits its sediments forming a cone shaped earth body. The course sediments like gravel and sand are deposited first, close to the entrance of the river into the plain. The finer sediments, consisting of silt and clay, are deposited towards the base of the cone. Upon entering the plain, and forced by the deposits of the sediments, the river divides itself into numerous branches fanning out towards the plain. The alluvial fans contain considerable groundwater reservoirs that are replenished each year by infiltration of the water from the river branches into the usually permeable underground thus obtaining rich aquifers.

The mountainous areas usually receive more rainfall than the plains: they form a watershed and provide a source of water. In (semi) arid regions, therefore, alluvial fans are often used for irrigation of agricultural crops. The fans reveal much greenery in the harsh desert-like environment. The irrigation methods in alluvial fans differ according to the hydrological regime of the river, the shape of the fan, and the natural resources available to maintain human life.

Types of Fans

The following alluvial fans will be reviewed in increasing order of water yield:

- The alluvial fans along the river plains near Khuzdar, Baluchistan, Pakistan. These fans are fed by small water catchments in areas of relatively low mountains. The fans are relatively small, steep, and subject to flash floods

- The alluvial fan of Garmsar, east of Tehran, Iran. This fairly large fan is fed by the *Hableh Rud* (river) with an important catchment area in the high Alburz mountain range. The river carries a large amount of water during the rainy season, otherwise the discharge is low.
- The alluvial fan of Punata in the *Valle Alto*, east of Cochabamba, Bolivia. The fan is fed by The *Rio Paracaya* river with a higher average discharge than the *Hableh Rud* river. Consequently the fan is fairly flat.
- The Okavango inland delta, near Maun, Botswana.

The delta receives an enormous amount of river inflow from Angola. Hence, the fan is so large and flat that it is rather called a delta. It takes six months for the peak inflow at the apex to reach the base of the delta.

Case Studies

Khuzdar

Average annual rainfall in Baluchistan varies between 200 and 400 mm, depending on altitude, and the main part occurs in winter (November to March). Of old, in sloping lands, farmers constructed bunds along the contour lines to capture the surface runoff. This method of water harvesting (locally called *khuskaba*) provided extra water for agricultural crops planted just up-slope of the bund, where the captured water would infiltrate into the soil and provide extra soil moisture to supplement the scarce rainfall.

In alluvial fans, the spate floods provided and additional source of water. The floods, diverted from the watercourses, were retained behind similar bunds employed in the *khuskaba* system. The method of flood interception is locally called *sailaba*. The system is combined with the tapping of groundwater from the aquifer by means of dug underground galleries, called *karez* or qanat. The *karezes* make permanent agriculture possible.

Although the *sailaba*-and-*qanat* systems cover about 20% of the agricultural land, their production is more than 40% of the total.

It is a modern development to sink deep wells in the aquifer of the alluvial fan to exploit the groundwater more effectively than do the traditional *karezes*. The owners of the wells may be entrepreneurs from elsewhere and the original population runs the risk of loosing the *karez* water when the wells draw the water table down to a deeper depth than that of the *karezes* so that the these fall dry.

Garmsar

The irrigation system for the Garmsar alluvial fan is quite well developed, to the extent that lined canals have been made and a large belt-canal crosses the fan through its middle. Roughly, the cropped area occupies 30% of the land each season, while 70% is left fallow. The winter crops are mainly wheat and barley, while the summer crops are cotton and melons. However, the planting of the new crops is done before harvesting the previous crops. Thus, there is a period of overlap during which 60% of the land is under crops. The fallow land is continuously rotated throughout the years, so that there exists no permanent fallow land, except along the fringes at the base of the fan where soil salinization occurs.

Cumulative frequency distribution of the annual average river discharge, showing a large variation. An estimated average annual water balance. It is seen that the storage of irrigation losses in the aquifer plays an important role. In the dry season the groundwater is used for irrigation by pumping from deep wells. A cross-section of the groundwater situation. The water rights are expressed in *sang*, a measure of continuous flow of about 10 l/s, but in practice it varies from 10 to more than 15 l/s. The water is delivered to about 100 tertiary units (often a village), within which the water is distributed by 12-day rotations amongst the farmers who each are entitled to receive the authorized *sangs* for a fixed number of hours during each rotation period. The village communities are, at the same time, water-user associations who take care of the water-distribution within the tertiary unit and they maintain the tertiary canals.

At present, the distribution of surface irrigation water to the villages is determined by the Garmsar Water Authority on the basis of the water rights and verbal agreements and communications with the water users in the absence of a written manual. The authority also maintains the irrigation canals and structures. The structures are sometimes re-designed to adjust them to verbally communicated needs. The fair distribution of the irrigation water is not an easy job as the average annual river discharge is quite variable in the range of 5 to 20 m3/s. The deep tube-wells are privately owned. The drilling of wells is subject to license. Recently, the licensing has stopped for fear of over-exploitation of the aquifer. It appears that no operational rules are applied to the wells. In the fringe lands, the water table is shallow because the discharge capacity of the aquifer diminishes here for two reasons: (1) the hydraulic gradient reduces where the sloping alluvial fan reaches the flat desert area, and (2) the thickness and hydraulic conductivity of the aquifer diminishes. The necessary drainage canals for watertable control at the fringes of the irrigation perimeter are not maintained by the water authority, but by the respective farmers groups. For the irrigation water, these groups depend (1) on occasional river floods too large to be handled by the irrigation system and that flow down to the fringe lands through the natural watercourses, (2) on spillage from the irrigation system, and (3) on deep wells.

To stabilize the agriculture in the fringe lands, which are threatened by soil salinization, a method of strip-cropping can be recommended for soil salinity control. This method uses irrigated strips next to permanently un-irrigated strips, whereby the salinization is directed to the un-irrigated strips. This concept is sometimes called *sacrificial drainage.*

Punata

The alluvial fan of Punata is found in the district of Cochabamba, Bolivia. The region of Punata, at the upper end of the Valle (valley) Alto, at about 2800 m altitude, has a summer rainfall of 400 to 450 mm starting in the second half of November end ending in March. Maize is here the most important food crop, followed by potatoes. Alfalfa is the dominant fodder crop, followed by maize straw. These crops could, of old, only be planted successfully because of the existence of additional water resources like runoff, floods, river base-flow and groundwater. In the winter months, crop growth is restricted due to the occurrence of night frosts, especially in June and July, and absence of rains. The total rural population in Punata is estimated at 25 000. There are about 4000 families of which an estimated 3680 are farmer families. The farms are small. The average size is 1.3 ha of which 1 ha is cropped. The modal size of farm is smaller, about 0.7 ha.

The rainfall distribution in Punata is characterized by a wet season from December to March, a dry season from May to October, and transition months in April and November. The average yearly total is 428 mm (1966 to 1983, San Benito). The rainfall with a probability of exceedance of 75% (R75) on a year basis is 360 mm. Rainfall is not reliable: in the period from 1966 to 1983, the yearly total varied between 246 mm (1982/83) to 591 mm (1968/69).

The river floods during the rainy summer period can be used for irrigation by anyone who wants to. When the river flow recedes, the stream can only be used for rotational irrigation by those who are entitled to take part in it (this is locally called the *mita* system). By the month of May the river base-flow becomes strongly reduced, and a drought period sets in, lasting into November. Irrigation is considered desirable to start the cropping season in August/September, so that an early harvest can be obtained. The early harvest has a high market value and reduces peak labour requirements. Further, the irrigation reduces the risk of crop failure and it permits diversification of agricultural produce. Nevertheless, there are some farming communities that have refrained in the past from the extra effort to obtain additional irrigation water and who seemed to be content with purely rain-fed cropping. At a modest scale, irrigation from deep-wells is also practiced.

In order to satisfy the needs of the majority of the farmers who strongly wish to have additional irrigation water, the irrigation project Punata-Tiraque began to be developed from 1970 onwards. The project entailed the construction of a complicated system of dams and reservoirs up in the Andean mountains.

The gross area of the Punata projects is estimated at 4600 ha, 90% of which can be used for agriculture or animal husbandry. About 1150 ha of this presently receive irrigation water, either surface water derived from the *Laguna Robada* or *Lluska Kocha* dam, or water pumped from the 16 deep wells in the project area (estimated at 350 ha). In addition there are a few hundred hectares that receive occasional water from *mita* irrigation (wild flooding).

The traditional irrigation method is based on handling large irrigation flows (*golpes*) per farm at large intervals. The intake structures in the *Pucara Mayu* river, at the place where it enters the alluvial fan of Punata, would alternately pass water from each of the reservoir systems (*Laguna Robada* and *Lluska Kocha/Muyu Loma*) and the natural *mita* water. The new system has been designed for smaller flows with shorter rotation intervals, but it works continuously for the whole area, so that there is no need anymore to separate the various sources of water. It covers a much larger area than the traditional system and it incorporates the associations of the *mita* systems (which may have partly the same members), the associations of tube-well systems (which may also have partly the same members) as well as the persons who had no previous water rights.

Hence, the new irrigation system makes it necessary to replace the traditional water rights by a totally new set of rights (and duties). In addition, the farmers will have to get used to new water distribution methods and new field irrigation techniques. Because the new irrigation zones do not correspond to the boundaries of the existing, scattered, *Comité's de Riego*, not only the water management but also the organizational structure will have to be adjusted to the new situation.

Okavango

Features

The inland Okavango Delta in northwest Botswana is hand-shaped, with the fingers spread. The Okavango River, which originates in Angola, enters the delta at its apex. On the average, the river carries about 10 000 million m^3 of water a year into the delta. The flow rate is high in the months of March and April (about 1000 m^3/s on the average), but varying from year to year between 500 to 1500 m^3/s and low in November (100 to 200 m^3/s).

The large volume of water spreading over the delta is almost fully absorbed in permanent and seasonal swamps (the latter are called *molapo's*) before is slowly infiltrates or evaporates. There is rich swamp vegetation, which creates an ideal environment for numerous kinds of animals. The rich fauna finds its habitat on and between the thousands of islands between the swamps.

The little water that exceeds the retention capacity of the marshy wilderness drains from July to November through the fingers of the giant hand. Hence, it takes almost six months before the peak discharge of the Okavango River manifests itself at the base of the delta. Here, the water meets a barrier: the Thamalakane Fault Line, beyond which the Kalahari sands rise up 10 m. At the foot of the fault, the Thamalakane River collects the water (which is not more than 5% of the total inflow) and carries it almost without gradient to the Boteti River, which flows through a breach in the fault line. Eventually, the remaining waters evaporate in the Makgadikgadi Pans, more than 200 km to the east.

Although the annual rainfall is relatively low (an average of 500 mm, the greater part of which falls from December to March), it contributes a volume of water to the Delta equalling half of the Okavango inflow. The annual rainfall and its yearly distribution are equally erratic as the regime of the river.

The Okavango River transports a large amount of sands and other sediments into the delta. Their mass is about 2 million tons a year. Salts also enter the Delta, but they do so in a dissolved form. The salt concentration of the water is some 200 mg/l, which is very low. The total weight of incoming salts is thus about 2 million tons per year.

The sediments and the salts imported by the Okavango River settle in the delta. Together with the vegetation the sediments build up resistances to surface-water flow. As a result, the main watercourses have in the past swayed from thumb to little finger to and fro, as is common in alluvial fans. Tectonic movements have also contributed to this phenomenon. At present the middle finger, from which the Boro River stems, provides the major thoroughfare.

Many of the islands in the delta have a garland of riparian trees along their borders, but in the middle they are bare: symptoms of salt accumulation.

The Kalahari Desert cooperates with the Okavango River to form the predominantly sandy soils of the delta. The desert uses the vector wind to deposit its share of fine sand.

The geophysical characteristics of the Okavango Delta have led to a low population density, so that the natural situation has scarcely been disrupted by mankind. Also, the population was more interested in hunting and cattle breeding than in food crop production, so that agricultural developments were limited. The arable lands in the south-eastern fringes of the delta, that become dry after the floods recede (these are locally known as molapo's), often have a sandy topsoil. In depressions, the topsoil may be thin or missing altogether, exposing a heavy clay soil.

Developments

In 1978/79, after four years of high and prolonged floods had made *molapo* farming impossible, a severe drought coincided with an outbreak of foot and mouth disease, leaving the local population in a state of emergency. This resulted in two important undertakings:

Hydrographs of the flood level of the Boro river, with an indication of the years in which the timely closure of the sluice gates would facilitate flood-recession cropping in the molapo's;

- The first took place from 1979 to 1981 when, as part of a drought-relief program (Food for Work), FAO organized labour-intensive works to rehabilitate the flood-control bunds that the local population had built to protect their crops against inundations from rainstorms. Some new bunds were also built.

- The second was the construction of the "Buffalo-fence" which separates the outer fringes of the delta from its interior to prevent the spread of cattle diseases – especially foot and mouth disease. Completed in 1983, this fence has increased the importance of the *molapo's* outside it, and for the following reasons. Especially in the years when rains start late, the new grass flush in the *molapo's* after flood-recession presents about the only source of fodder in the region, also for the herds in the wide surroundings of the delta. With the *molapo's* inside the fence closed to cattle grazing, the grazing intensity in the *molapo's* outside the fence has increased.

The Molapo Development Projects (MDP) became operational in December 1983. The project aimed at increasing crop production in pilot areas by protecting those areas against prolonged high floods by flood-control bunds with entrance gates that could be closed. When enough floodwater has entered the molapo, the gate can be closed and the water recedes under influence of evaporation and infiltration into the soil and cropping may start when the outside water level is still high. This was a response to the high and prolonged flooding in the years 1974-1978, when *molapo* cropping was largely impossible. In more recent years, however, inadequate floods appeared to present an equally severe constraint to satisfactory crop production. It was therefore decided to focus also on improved, more stable crop production under fully rain-dependent conditions.

After the flood has been permitted to enter the bunded molapo, the sluice gates are closed. Recession of the water in the bunded molapo then begins under the influence of evaporation and infiltration, and allows a timely planting of the crop (in October or November). The crops use the residual soil moisture (about 100 mm) till the onset of the rainy season at the end of November or the beginning of December. Thus the growing season is prolonged, the moisture availability is increased, and crop production is enhanced. However, the success of the flood-control measures on the crop performance still depends to a great extent on the amount and distribution of the rainfall.

Drainage

Drainage is the natural or artificial removal of surface and subsurface water from an area. Many agricultural soils need drainage to improve production or to manage water supplies.

Early History

The earliest archaeological record of an advanced system of drainage comes from the Indus Valley Civilization from around 3100 BC in what is now Pakistan and North India. The ancient Indus systems of sewerage and drainage that were developed and used in cities throughout the civilization were far more advanced than any found in contemporary urban sites in the Middle East and even more efficient than those in some areas of modern Pakistan and India today. All houses in the major cities of Harappa and Mohenjodaro had access to water and drainage facilities. Waste water was directed to covered drains, which lined the major streets.

Drainage in the 19th Century

This operation is always best performed in spring or summer, when the ground is dry. Main drains ought to be made in every part of the field where a crosscut or open drain was formerly wanted; they ought to be cut four feet (1.2 m) deep, upon an average. This completely secures them from the possibility of being damaged by the treading of horses or cattle, and being so far below the small drains, clears the water finely out of them. In every situation, pipe-turfs for the main drains, if they can be had, are preferable. If good stiff clay, a single row of pipe-turf; if sandy, a double row. When pipe-turf cannot be got conveniently, a good wedge drain may answer well, when the subsoil is a strong, stiff clay; but if the subsoil be only moderately so, a thorn drain, with couples below, will do still better; and if the subsoil is very sandy, except pipes can be had, it is in vain to attempt under-draining the field by any other method. It may be necessary to mention here that the size of the main drains ought to be regulated according to the length and declivity of the run, and the quantity of water to be carried off by them. It is always safe, however, to have the main drains large, and plenty of them; for economy here seldom turns out well.

Having finished the main drains, proceed next to make a small drain in every furrow of the field if the ridges formerly have not been less than fifteen feet (5 m) wide. But if that should be the case, first level the ridges, and make the drains in the best direction, and at such a distance from each other as may be thought necessary. If the water rises well in the bottom of the drains, they ought to be cut three feet (1 m) deep, and in this ease would dry the field sufficiently well, although they were from twenty-five to thirty feet (8 to 10 m) asunder; but if the water does not draw well to the bottom of the drains, two feet (0.6 m) will be a sufficient deepness for the pipe-drain, and two and a half feet (1 m) for the wedge drain. In no case ought they to be shallower where the field has been previously levelled. In this instance, however, as the surface water is carried off chiefly by the water sinking immediately into the top of the drains, it will be necessary to have the drains much nearer each other—say from fifteen to twenty feet (5 to 6 m). If the ridges are more than fifteen feet (5 m) wide, however broad and irregular they may be, follow invariably the line of the old furrows, as the best direction for the drains; and, where they are high-gathered ridges, from twenty to twentyfour inches will be a sufficient depth for the pipe-drain, and from twentyfour to thirty inches for the wedge-drain. Particular care should be taken in connecting the small and main drains together, so that the water may have a gentle declivity, with free access into the main drains.

When the drains are finished, the ridges are cleaved down upon the drains by the plough; and where they had been very high formerly, a second clearing may be given; but it is better not to level the ridges too much, for by allowing them to retain a little of their former shape, the ground being lowest immediately where the drains are, the surface water collects upon the top of the drains; and, by shrinking into them, gets freely away. After the field is thus finished, run the new ridges across the small drains, making them about ten feet (3 m) broad, and continue afterwards to plough the field in the same manner as dry land. It is evident from the above method of draining that the expense will vary very much, according to the quantity of main drains necessary for the field, the distance of the small drains from each other, and the distance the turf is to be carried.

The advantage resulting from under-draining, is very great, for besides a considerable saving annually of water furrowing, cross cutting, etc., the land can often be ploughed and sown to advantage, both in the spring and in the fall of the year, when otherwise it would be found quite impracticable; every species of drilled crops, such as beans, potatoes, turnips, etc., can be cultivated successfully; and every species, both of green and white crops, is less apt to fail in wet and untoward seasons. Wherever a burst of water appears in any particular spot, the sure and certain way of getting quit of such an evil is to dig hollow drains to such a depth below the surface as is required by the fall or level that can be gained, and by the quantity of water expected to proceed from the burst or spring. Having ascertained the extent of water to be carried off, taken the necessary levels, and cleared a mouth or loading passage for the water, begin the drain at the extremity next to that leader, and go on with the work till the top of the spring is touched, which probably will accomplish the intended object. But if it should not be completely accomplished, run off from the main drain with such a number of branches as may be required to intercept the water, and in this way disappointment will hardly be experienced. Drains, to be substantially useful, should seldom be less than three feet (1 m) in depth, twenty or twenty four inches thereof to be close packed with stones or wood, according to circumstances. The former are the best materials, but in many places are not to be got in sufficient quantities; recourse therefore, must often be made to the latter, though not so effectual or durable.

It is of vast importance to fill up drains as fast as they are dug out; because, if left open for any length of time, the earth is not only apt to fall in but the sides get into a broken, irregular state, which cannot afterwards be completely rectified. A proper covering of straw or sod should be put upon the top of the materials, to keep the surface earth from mixing with them; and where wood is the material used for filling up, a double degree of attention is necessary, otherwise the proposed improvement may be effectually frustrated.

The pit method of draining is a very effectual one, if executed with judgment. When it is sufficiently ascertained where the bed of water is deposited, which can easily be done by boring with an auger, sink a pit into the place of a size which will allow a man freely to work within its bounds. Dig this pit of such a depth as to reach the bed of the water meant to be carried off; and when this depth is attained, which is easily discerned by the rising of the water, fill up the pit with great land-stones and carry off the water by a stout drain to some adjoining ditch or mouth, whence it may proceed to the nearest river.

Current Practices

Modern drainage systems incorporate geotextile filters that retain and prevent fine grains of soil from passing into and clogging the drain. Geotextiles are synthetic textile fabrics specially manufactured for civil and environmental engineering applications. Geotextiles are designed to retain fine soil particles while allowing water to pass through. In a typical drainage system they would be laid along a trench which would then be filled with coarse granular material: gravel, sea shells, stone or rock. The geotextile is then folded over the top of the stone and the trench is then covered by soil. Groundwater seeps through the geotextile and flows within the stone to an outfall. In high groundwater conditions a perforated plastic (PVC or PE) pipe is laid along the base of the drain to increase the volume of water transported in the drain. Alternatively, prefabricated plastic drainage systems made of HDPE called Smart Ditch, often incorporating geotextile, coco fiber or rag filters can be considered. The use of these materials has become increasingly more common due to their ease of use which eliminates the need for transporting and laying stone drainage aggregate which is invariably more expensive than a synthetic drain and concrete liners. Over the past 30 years geotextile and PVC filters have become the most commonly used soil filter media. They are cheap to produce and easy to lay, with factory controlled properties that ensure long term filtration performance even in fine silty soil conditions.

21st Century Alternatives

Seattle's Public Utilities created a pilot program called Street Edge Alternatives (SEA Streets) Project. The project focuses on designing a system "to provide drainage that more closely mimics the natural landscape prior to development than traditional piped systems". The streets are characterized by ditches along the side of the roadway, with plantings designed throughout the area. An emphasis on non curbed sidewalks allows water to flow more freely into the areas of permeable surface on the side of the streets. Because of the plantings the run off water from the urban area does not all directly go into the ground but can also be absorbed into the surrounding environment. According to the monitoring by Seattle Public Utilities, they report at 99 percent reduction of storm water leaving the drainage project

Drainage in Construction

The civil engineer or site engineer is responsible for drainage in construction projects. They set out from the plans all the roads, Street gutters, drainage, culverts and sewers involved in construction operations. During the construction of the work on site he/she will set out all the necessary levels for each of the previously mentioned factors.

Site engineers work alongside architects and construction managers, supervisors, planners, quantity surveyors, the general workforce, as well as subcontractors. Typically, most jurisdictions have some body of drainage law to govern to what degree a landowner can alter the drainage from his parcel.

Reasons for Artificial Drainage

Wetland soils may need drainage to be used for agriculture. In the northern USA and Europe, glaciation created numerous small lakes which gradually filled with humus to make marshes. Some of these were drained using open ditches and trenches to make mucklands, which are primarily used for high value crops such as vegetables.

The largest project of this type in the world has been in process for centuries in the Netherlands. The area between Amsterdam, Haarlem and Leiden was, in prehistoric times swampland and small lakes. Turf cutting (Peat mining), subsidence and shoreline erosion gradually caused the formation of one large lake, the Haarlemmermeer, or lake of Haarlem. The invention of wind powered pumping engines in the 15th century permitted drainage of some of the marginal land, but the final drainage of the lake had to await the design of large, steam powered pumps and agreements between regional authorities. The elimination of the lake occurred between 1849 and 1852, creating thousands of km^2 of new land.

Coastal plains and river deltas may have seasonally or permanently high water tables and must have drainage improvements if they are to be used for agriculture. An example is the flatwoods citrus-growing region of Florida. After periods of high rainfall, drainage pumps are employed to prevent damage to the citrus groves from overly wet soils. Rice production requires complete control of water, as fields need to be flooded or drained at different stages of the crop cycle. The Netherlands has also led the way in this type of drainage, not only to drain lowland along the shore, but actually pushing back the sea until the original nation has been greatly enlarged. In moist climates, soils may be adequate for cropping with the exception that they become waterlogged for brief periods each year, from snow melt or from heavy rains. Soils that are predominantly clay will pass water very slowly downward, meanwhile plant roots suffocate because the excessive water around the roots eliminates air movement through the soil. Other soils may have an impervious layer of mineralized soil, called a hardpan or relatively impervious rock layers may underlie shallow soils. Drainage is especially important in tree fruit production. Soils that are otherwise excellent may be waterlogged for a week of the year, which is sufficient to kill fruit trees and cost the productivity of the land until replacements can be established. In each of these cases appropriate drainage carries off temporary flushes of water to prevent damage to annual or perennial crops.

Drier areas are often farmed by irrigation, and one would not consider drainage necessary. However, irrigation water always contains minerals and salts, which can be concentrated to toxic levels by evapotranspiration. Irrigated land may need periodic flushes with excessive irrigation water and drainage to control soil salinity.

Tile Drainage

Tile drainage (in agriculture) is a practice for removing excess water from the subsurface of soil intended for agriculture. Whereas irrigation is the practice of adding additional water when the soil is naturally too dry, drainage brings soil moisture levels down for optimal crop growth. While surface water can be drained via pumping and/ or open ditches, tile drainage is often the best recourse for subsurface water. Too much subsurface water can be counterproductive to agriculture by preventing root development, and inhibiting the growth of crops. Too much water also can limit access to the land, particularly by farm machinery. In terms of access, most modern agriculture depends on the usage of large machinery—tractors and implements— to prepare the seedbed, plant the crop, carry out any cultivation and applications during the growing season, and ultimately, to harvest the crop. Operating most machinery in excessively wet conditions may result in soil degradation because of excessive soil compaction, and inhibit the operation of the machinery (i.e., "getting stuck").

Increased Crop Yields

Most crops require specific soil moisture conditions, and do not grow well in wet, "mucky" soil. Even in soil that isn't "mucky" the roots of most plants will not grow much deeper than the water table. Early in the growing season when water is in ample supply, plants are small and don't require much water. During this time, the plants do not need to develop their roots to "reach" for the water. As the plants grow and use more water during the growing season water becomes more scarce. During this time, the water table starts to fall. Plants suddenly need to start developing roots to reach to the water. During dry times, the water table can fall faster than the plants can develop roots to "reach" for the water. This can seriously stress the plants.

By adding drain tile, the water table is effectively lowered, and plants can properly develop their roots. The lack of water saturation allows oxygen to exist in the soil around the roots. Drain tile prevents the roots from being under the water table during wet periods that could cause excessive plant stress. By removing excessive water, crops use water they have more effectively. An increase in crop yield can be simply summarized by the following: Simply by forcing the plants to have more developed roots, the plants can more effectively absorb more nutrients and water.

The same principal is employed by containers that hold house plants: they have drain holes in the bottom to allow oxygen to the roots, and prevent soil saturation. By placing drain tile under a field in a grid style layout, the same effect can be employed on a several hundred acre field.

Plumbing of Drain Tile

In a tile drainage system, a sort of "plumbing" is installed below the surface of agricultural fields, effectively consisting of a network of below-ground pipes that allow subsurface water to move out from between soil particles and into the tile line. Water flowing through tile lines is often ultimately deposited into surface water points—lakes, streams, and rivers—located at a lower elevation than the source. Water enters the tile line either via the gaps between tile sections, in the case of older tile designs, or through small perforations in modern plastic tile. Soil type greatly affects the efficacy of tile systems, and dictates the extent to which the area must be tiled in order to ensure sufficient drainage. Sandier soils will need little, if any, additional drainage, whereas soils with high clay contents will hold their water tighter, requiring tile lines to be placed closer together.

History of Tile Drainage

Both Cato and Pliny have described tile drainage systems, in 200 BC and the first century AD, respectively. According to the Johnston Farm website, tile drainage was first introduced to the United States in 1838, when John Johnston brought the practice from his native Scotland to his farm in Seneca County, New York. Johnston laboured to lay 72 miles (116 km) worth of clay tile on 320 acres (1.3 km^2). The effort paid off by increasing his wheat yield from 12 bushels per acre to 60 bu/acre. Johnston, "Father of American Tile Drainage", continued to advocate tile drainage throughout his life, attributing his success as a farmer to the formula "D,C, and D" (dung, credit, and drainage).

The expansion of drainage networks was an important technical aspect of Westward Expansion in the 19th century. Although land in the United States was parcelled out in accordance with the Public Land Survey System as established by the Land Ordinance of 1785, development, especially of agricultural land, was often limited by the rate at which it was made capable for cultivation. For example, although Iowa was made a state in 1846, maps depicting land ownership show below-average population densities in the northwestern region as late as the 1870s, a corner of the state which today is still noted for its high water table and numerous lakes, marshes, and wetlands.

States throughout the region faced similar limits to agricultural intensification. Many states offered government incentives to improve land for farming. For example, legislation in Indiana prompted an Act of Congress in 1850 that provided for swamplands to be sold at a discount to farmers on the condition that they drain the land and bring it into agricultural productivity. To facilitate this process, most states set up government agencies to oversee and regulate the installation of tile drainage systems. Even today, ballots for elections in rural America often include candidates for local drainage supervisory boards. The decades following the American Civil War saw rapid expansion of drainage systems. For example, historical literature from Ohio notes that in the year of 1882, the number of acres drained was about equal to the area of land drained in all previous years. In the 1930s, the Civilian Conservation Corps contributed to the tile network throughout the Midwest, much of which is still in use.

Advances in Drainage Technology

Throughout the twentieth century, the technology of tile installation remained similar to the methods first used in 1838. Although cement sections later replaced the original clay tiles, and machines were used to dig the trenches for the tile lines, the process remained quite labour-intensive and limited to specialized contractors.

The introduction of plastic tile served to reduce both the cost of tile installation, as well as the amount of labour involved. Rather than set individual sections of cement tile end-to-end in the trench, tile installers had only to unroll a continuous section of lightweight, flexible tile line. Towards the end of the twentieth century, when large, four-wheel-drive tractors became more common on American farms, do-it-yourself tile implements appeared on the market. By making tile installation cheaper and allowing it to be done on the landowner's schedule, farmers are capable of draining localized wet spots that may not create enough of a problem to merit more costly operations. In this way, farmers may enjoy increases in crop yield while saving on the capital costs of tile installation.

Social and Ecological Effects of Tile Drainage

Unfortunately, the ability for farmers to install their own tile can be problematic. First of all, private installations may reduce the ability of local drainage supervisory boards to regulate tile installation, which in some areas of the country requires proper documentation before a contractor can continue. This leads to the second potential conflict, the unintentional interruption of existing tile networks. Most "do-it-yourself" tile flows do not dig trenches but rather split the soil enough to squeeze the tile line in; thus, a farmer would not be aware if he breaks a line of tile that might serve his neighbors, as well (Mutual tile lines are often dictated by topography rather than land ownership, and the location of many old, but still effective, tile lines are unknown.). The potential for across-the-fence disputes are obvious.

Ecologically, the expansion of drainage systems has had tremendous negative effects. Hundreds of thousands of wetland species experienced significant population declines as their habitat was increasingly fragmented and destroyed. Although market hunting within the Central Flyway was a contributing factor in the decline of many waterfowl species' numbers in the early decades of the twentieth century, loss of breeding habitat to agricultural expansion is certainly the most significant. Early maps of midwestern states depict many lakes and marshes that are either nonexistent or significantly reduced in area today. Channelization, a related process of concentrating and faciliting the flow of water from agricultural areas, also contributed to this degradation.

In bypassing the natural flow of water from the surface to the water table, drainage systems often prevent the natural filtration of water provided by soils and wetlands. Thus, drainage systems pose a threat to surface water sources by directly depositing water laden with fertilizers, eroded soil, agrochemicals, and other types of agricultural runoff pollutants. In very flat areas, where the natural topography does not provide the gradient necessary for water flow, "agricultural wells" can be dug to provide tile lines sufficient outlet. In these cases, it is the groundwater that stands to be polluted by unfiltered tile output.

Intensive Livestock Operations (ILO) have led to challenges of livestock effluent disposal. Livestock effluent contains valuabe nutrients, but the misapplication of these materials can lead to serious ecological problems, such as nutrient loading. Injecting effluent directly into the ground is one method employed by manure applicators to improve nutrient uptake. Drainage tiles may increase injected manure seepage into surface waterways from manure injection because liquid manure seeps through soils and then drains out of the field and into waterways via drainage tiles.

Today, a number of state and federal initiatives serve to reverse habitat loss. Many programs encourage and even reimburse farmers for interrupting the drainage of localized wetholes on their property, often by breaking tile intakes or removing the tile completely.

Landowners are often partially or fully compensated for forfeiting the ability to grow crops on this land. Such programs and the cooperation of landowners across the country have had significant positive effects on the populations of a wide variety of waterfowl.

Drainage System (Agriculture)

An agricultural drainage system is a system by which the water level on or in the soil is controlled to enhance agricultural crop production.

Classification

The function of the field drainage system is to control the water table, whereas the function of the main drainage system is to collect, transport, and dispose of the water through an outfall or outlet. In some instances one makes an additional distinction between collector and main drainage systems.

Field drainage systems are differentiated in surface and subsurface field drainage systems. Sometimes (e.g. in irrigated, submerged rice fields), a form of temporary drainage is required whereby the drainage system is allowed to function on certain occasions only (e.g. during the harvest period). If allowed to function continuously, excessive quantities of water would be lost. Such a system is therefore called a checked, or controlled, drainage system. More usually, however, the drainage system is meant to function as regularly as possible to prevent undue waterlogging at any time and one employs a regular drainage system. In literature, this is sometimes also called a "relief drainage system".

Surface Drainage Systems

The regular surface drainage systems, which start functioning as soon as there is an excess of rainfall or irrigation, operate entirely by gravity. They consist of reshaped or reformed land surfaces and can be divided into:

- Bedded systems, used in flat lands for crops other than rice;
- Graded systems, used in sloping land for crops other than rice.

The bedded and graded systems may have ridges and furrows.

The checked surface drainage systems consist of check gates placed in the embankments surrounding flat basins, such as those used for rice fields in flat lands. These fields are usually submerged and only need to be drained on certain occasions (e.g. at harvest time). Checked surface drainage systems are also found in terraced lands used for rice. In literature, not much information can be found on the relations between the various regular surface field drainage systems, the reduction in the degree of waterlogging, and the agricultural or environmental effects. It is therefore difficult to develop sound agricultural criteria for the regular surface field drainage systems. Most of the known criteria for these systems concern the efficiency of the techniques of land levelling and earthmoving.

Similarly, agricultural criteria for checked surface drainage systems are not very well known.

Subsurface Drainage Systems

Like the surface field drainage systems, the subsurface field drainage systems can also be differentiated in regular systems and checked (controlled) systems. When the drain discharge takes place entirely by gravity, both types of subsurface systems have much in common, except that the *checked* systems have control gates that can be opened and closed according to need. They can save much irrigation water. A *checked* drainage system also reduces the discharge through the main drainage system, thereby reducing construction costs.

When the discharge takes place by *pumping*, the drainage can be checked simply by not operating the pumps or by reducing the pumping time. In northwestern India, this practice has increased the irrigation efficiency and reduced the quantity of irrigation water needed, and has not led to any undue salinization. The subsurface field drainage systems consist of horizontal or slightly sloping channels made in the soil; they can be open ditches, trenches, filled with brushwood and a soil cap, filled with stones and a soil cap, buried pipe drains, tile drains, or mole drains, but they can also consist of a series of wells. Modern buried pipe drains often consist of corrugated, flexible, and perforated plastic (PE or PVC) pipe lines wrapped with an *envelope* or filter material to improve the permeability around the pipes and to prevent entry of soil particles, which is especially important in fine sandy and silty soils. The surround may consist of synthetic fibre (geotextile).

The *field drains* (or *laterals*) discharge their water into the collector or main system either by *gravity* or by *pumping*. The wells (which may be open dug wells or *tubewells*) have normally to be pumped, but sometimes they are connected to drains for discharge by gravity. Subsurface drainage by wells is often referred to as vertical drainage, and drainage by channels as horizontal drainage, but it is more clear to speak of "field drainage by wells" and "field drainage by ditches or pipes" respectively.

In some instances, subsurface drainage can be achieved simply by breaking up slowly permeable soil layers by *deep plowing* (*sub-soiling*), provided that the underground has sufficient natural drainage. In other instances, a combination of sub-soiling and subsurface drains may solve the problem.

Main Drainage Systems

The main drainage systems consist of *deep* or *shallow* collectors, and *main drains* or *disposal drains*. Deep collector drains are required for subsurface field drainage systems, whereas shallow collector drains are used for surface field drainage systems, but they can also be used for pumped subsurface systems. The deep collectors may consist of open ditches or buried pipe lines. The terms *deep collectors* and *shallow collectors* refer rather to the depth of the water level in the collector below the soil surface than to the depth of the bottom of the collector. The bottom depth is determined both by the depth of the water level and by the required discharge capacity.

The deep collectors may either discharge their water into deep main drains (which are drains that do not receive water directly from field drains, but only from collectors), or their water may be pumped into a disposal drain. *Disposal* drains are main drains in which the depth of the water level below the soil surface is not bound to a minimum, and the water level may even be above the soil surface, provided that embankments are made to prevent inundation. Disposal drains can serve both subsurface and surface field drainage systems.

Deep main drains can gradually become disposal drains if they are given a smaller gradient than the land slope along the drain. The technical criteria applicable to main drainage systems depend on the hydrological situation and on the type of system.

Main Drainage Outlet

The final point of a main drainage system is the gravity outlet structure or the pumping station.

Applications

Surface drainage systems are usually applied in relatively flat lands that have soils with a low or medium infiltration capacity, or in lands with high-intensity rainfalls that exceed the normal infiltration capacity, so that frequent waterlogging occurs on the soil surface.

Subsurface drainage systems are used when the drainage problem is mainly that of shallow water tables.

When both surface and subsurface waterlogging occur, a combined surface/subsurface drainage system is required. Sometimes, a subsurface drainage system is installed in soils with a low infiltration capacity, where a surface drainage problem may improve the soil structure and the infiltration capacity so greatly that a surface drainage system is no longer required. On the other hand, it can also happen that a surface drainage system diminishes the recharge of the groundwater to such an extent that the subsurface drainage problem is considerably reduced or even eliminated.

The choice between a subsurface drainage system by pipes and ditches or by tube wells is more a matter of technical criteria and costs than of agricultural criteria, because both types of systems can be designed to meet the same agricultural criteria and achieve the same benefits. Usually, pipe drains or ditches are preferable to wells. However, when the soil consists of a poorly permeable top layer several meters thick, overlying a rapidly permeable and deep subsoil, wells may be a better option, because the drain spacing required for pipes or ditches would be considerably smaller than the spacing for wells. When the land needs a subsurface drainage system, but saline groundwater is present at great depth, it is better to employ a shallow, closely spaced system of pipes or ditches instead of a deep, widely spaced system. The reason is that the deeper systems produce a more salty effluent than the shallow systems. Environmental criteria may then prohibit the use of the deeper systems.

In some drainage projects, one may find that only main drainage systems are envisaged. The agricultural land is then still likely to suffer from field drainage problems. In other cases, one may find that field drainage systems are ineffective because there is no adequate main drainage system. In either case, the installation of drainage systems is not recommended. Reference: gives a general description of land drainage in the world and shows a paper on types of agricultural land drainage systems used in different parts of the world.

Drainage System Design

The analysis of positive and negative (side) effects of drainage and the optimization of drainage design in accordance to the *drainage design procedures* is discussed in the article on Drainage research.

Watertable Control

Watertable control is the practice of controlling the water table in agricultural land by subsurface drainage with proper criteria to improve the crop production.

Description and Definitions

Subsurface land drainage aims at controlling the water table of the groundwater in originally waterlogged land at a depth acceptable for the purpose for which the land is used. The depth of the water table with drainage is *greater* than without.

Purpose

In agricultural land drainage, the purpose of water table control is to establish a depth of the water table that does no longer interfere negatively with the necessary farm operations and crop yields. In addition, land drainage can help with soil salinity control. The soil's hydraulic conductivity plays an important role in drainage design.

Optimization

Optimization of the depth of the water table is related the benefits and costs of the drainage system. The shallower the permissible depth of the water, the lower the cost of the drainage system to be installed to achieve this depth. However, the lowering of the originally too shallow depth by land drainage entails *side effects*. These have also to be taken into account, including the costs of mitigation of negative side effects. The *optimization* of drainage design and the development of *drainage criteria* are discussed in the article on drainage research. Example of the effect of drain depth on soil salinity and various irrigation/drainage parameters as simulated by the Salt Mod program.

History

Historically, agricultural land drainage started with the digging of relatively shallow open ditches that received both runoff from the land surface and outflow of groundwater. Hence the ditches had a surface as well as a subsurface drainage function. By the end of the 19th century and early in the 20th century it was felt that the ditches were a hindrance for the farm operations and the ditches were replaced by buried lines of clay pipes (tiles), each tile about 30 cm long. Hence the term "tile drainage". Since 1960, one started using long, flexible, corrugated plastic (PVC or PE) pipes that could be installed efficiently in one go by trenching machines. The pipes could be pre-wrapped with an envelope material, like synthetic fibre and geotextile, that would prevent the entry of soil particles into the drains.

Thus, land drainage became a powerful industry. At the same time agriculture was steering towards maximum productivity, so that the installation of drainage systems came in full swing.

Environment

As a result of large scale developments, many modern drainage projects were *over-designed*, while the negative environmental side effects were ignored. In circles with environmental concern, the profession of land drainage got a poor reputation, sometimes justly so, sometimes unjustified, notably when land drainage was confused with the more encompassing activity of wetland reclamation. Nowadays, in some countries, the hardliner trend is reversed. Further, *checked* or *controlled* drainage systems were introduced, as shown in discussed on the page: Drainage system (agriculture).

Drainage Design

The design of subsurface drainage systems in terms layout, depth and spacing of the drains is often done using subsurface drainage equations with parameters like drain depth, depth of the water, soil depth, hydraulic conductivity of the soil and drain discharge. The drain discharge is found from an agricultural water balance.

Drainage by Wells

Subsurface drainage of groundwater can also be accomplished by pumped wells (*vertical* drainage, in contrast to *horizontal* drainage). Drainage wells have been used extensively in the Salinity Control and Reclamation Program (SCARP) in the Indus valley of Pakistan. Although the experiences were not overly successful, the feasibility of this technique in areas with deep and permeable aquifers is not to be discarded. The well spacings in these areas can be so wide (more than 1000m) that the installation of *vertical* drainage systems could be relatively cheap compared to *horizontal* subsurface drainage (drainage by pipes, ditches, trenches, at a spacing of 100m or less). For the design of a well field for control of the water, the Well Drain model may be helpful.

3

River Water Management

The plight of the Snake River salmon runs has become one of the late twentieth century's foremost environmental issues. The Snake River is the largest tributary of the Columbia River, and the Snake Basin, including tributaries such as Idaho's famed Salmon River, once contained the most productive salmon spawning areas in the Columbia Basin. All Snake River salmon must traverse the lower 140 miles of the Snake River in eastern Washington. Unfortunately, the lower Snake, the gateway to all of Idaho's prime salmon habitat, is no longer a free-flowing river but is instead a series of reservoirs formed by four federal dams constructed and operated by the U.S. Army Corps of Engineers (Corps).

These Snake River dams, together with four additional Corps dams on the lower Columbia, all part of the federal Columbia River Power System, pose substantial hurdles to both upstream and downstream migrating salmon and have destroyed important spawning grounds. Although there are other causes for the decline of Snake River salmon, the principal factor leading to the decline and subsequent listing of the runs under the protection of the Endangered Species Act (ESA) was the construction and operation of these dams. We believe that as long as the Snake River salmon runs continue their plunge toward extinction, federal agencies, downstream economic and environmental interests, and Indian tribes with treaty-reserved fishing rights will continue to exert substantial pressure on upper Snake River Basin reservoir operators to release large amounts of water to augment fish flows in the lower Snake and Columbia Rivers. Release of large amounts of water stored in upper Snake River reservoirs could substantially affect Idaho farmers who rely on this water for irrigation, as well as the state's primary electric utility, Idaho Power Company, which relies on this water to generate hydropower. The region must weigh these impacts against those of the only other serious option for recovering the decimated Snake River fish runs—dam breaching and reservoir drawdowns. Mounting biological evidence indicates that breaching the four lower Snake dams and drawing down John Day Reservoir (on the Columbia River) offers a far better likelihood of recovering Idaho's salmon.

Moreover, this strategy would allow Idaho to continue to make use of Snake River Basin water for agriculture and power, thereby strengthening its economy. For these reasons, we contend that this alternative makes the most scientific, economic, and political sense for the state of Idaho, the Pacific Northwest, and the nation. Even before Snake River salmon were listed under the ESA in the 1990s, the Northwest states had implemented a restoration program under the terms of the Northwest Power Act. Despite the expenditure of considerable sums of money, however, the states' restoration program was unable to reverse the decline of Snake River salmon, a fact that the chairman of the interstate agency charged with implementing the program, the Northwest Power Planning Council (Council), recognized on the eve of the ESA listings.

The ESA listings led to the institution of what amounts to an interim federal recovery plan for Snake River salmon, which has governed dam operations since 1992. Despite the ESA's reputation as an economically insensitive, preservation-at-all-costs statute, in the section 7 biological opinion that today governs federal dam operations, the National Marine Fisheries Service (NMFS) authorized dam-related salmon mortality of up to eighty-six percent of juvenile sockeye and spring/summer chinook and up to ninety-nine percent of juvenile fall chinook. Although NMFS recognized the need for "major modifications" to the dams, no significant changes have yet occurred. Instead, the current centerpiece of federal agencies' interim salmon restoration efforts continues to be a program of barging and trucking juvenile fish around the dams, a technique that has failed to survive scientific scrutiny. Not surprisingly, federal restoration attempts show no signs of recovering the listed salmon species. Recognizing the inadequacy of current measures, NMFS has promised to issue a new road map for reconfiguring the hydrosystem to allow for salmon recovery in late 1999, when its current biological opinion on Columbia Basin hydroelectric operations expires. One idea that has gained prominence is permanent reservoir drawdowns—breaching the earthen portions of several dams—to allow restoration of natural river flows.

In 1995, the Indian tribes of the Nez Perce, Umatilla, Warm Springs, and Yakama reservation, which possess treaty fishing rights to Columbia Basin salmon, released their own restoration plan. The tribal plan included long-term permanent drawdowns of the four lower Snake reservoirs and the John Day Reservoir on the lower Columbia to achieve natural river flows. A year later, in 1996, the Council's group of independent scientists issued a report that lent support to the tribes' emphasis on natural river flows by calling for the re-establishment of what the report termed "normative river conditions." The scientists' most prominent recommendation was to restore river habitat by permanently lowering two large reservoirs on the lower Columbia, John Day and McNary, in order to revitalize drowned alluvial reaches and re-establish salmon spawning. The scientists also suggested that lowering the Snake River reservoirs to re-establish natural river flows would be consistent with the report's normative river concept.

Predictably, the discussions concerning drawing down lower Snake River reservoirs to achieve natural river flows generated a wave of political reaction. No Northwest politician endorsed the proposal, and several reacted with hostility, rather than carefully examining it. Senator Slade Gorton (R-Wa.) even sponsored an appropriations rider that would prevent any changes in the operation of Columbia Basin dams absent congressional approval. Yet these dams are hardly the linchpin of the regional economy—the lower Snake River dams were authorized as makework projects, and they provide no flood control and only marginal hydropower, navigation, and irrigation benefits. Moreover, economic analyses indicate that the overall benefits of drawdowns would substantially outweigh their costs, especially for the state of Idaho. Without reservoir drawdowns on the lower Snake, pressure to draft upstream Idaho reservoirs to facilitate salmon migration is likely to increase, either to satisfy the requirements of the ESA, the Northwest Power Act, the Clean Water Act, or Indian treaty rights. The alternative to breaching the lower Snake River dams thus appears to be drafting substantial amounts of water from upper Snake River reservoirs, with potentially adverse effects on irrigated agriculture, power generation, and resident fish. Idaho could avoid the adverse economic consequences associated with contributing increasing amounts of water stored in Idaho reservoirs if natural river flows were restored by breaching the lower Snake River dams and drawing down John Day Reservoir.

Reservoir drawdowns and natural river flows could therefore save both Snake River water and salmon. The Idaho Department of Fish and Game (IDFG) recently concluded that a drawdown to natural river flows "is the best biological choice for recovering [Snake River salmon]." Permanent natural river drawdowns also could form the basis for settling substantial Indian water right claims that the Nez Perce Tribe and the federal government are litigating in Idaho's long-running Snake River Basin Adjudication, as well as satisfy the bulk of Idaho's ESA obligations for Snake River salmon.

The Condition of Snake River Salmon and the Status of Recovery Efforts

Once the mainstay of the Columbia Basin salmon runs, all of the Snake River salmon are now either extinct or threatened with extinction. Snake River coho were declared extinct in 1986. Sockeye were listed as endangered under the Endangered Species Act (ESA) in 1991 and continue to teeter on the brink of extinction, with few or no adults returning each year. Chinook were listed as threatened in 1992, and steelhead in 1997. Formerly the most important spawning area for fall chinook in the Columbia Basin through the 1950s with nearly 30,000 spawners, the Snake River runs now average around 900 fish. Snake River spring and summer chinook historically made up nearly one-half of all Columbia Basin spring and summer chinook, with roughly 125,000 adults in the late 1950s, but averaged fewer than 10,000 wild fish during the 1980s.

The situation grew so grave that the National Marine Fisheries Service (NMFS), which originally listed the chinook species as threatened in 1992, temporarily downgraded their condition to endangered just two years later. In 1998, NMFS added Snake River steelhead to the list of threatened species, citing a decline from an average of 70,000 through 1970 to fewer than 10,000 in the early 1990s. Although by the 1960s Snake River salmon runs were less than 10 percent of their predevelopment size, they still averaged over 100,000 adults, as compared with an average of about 20,000 wild and hatchery fish over the past decade, and a total run of barely 2000 fish in 1995. This enormous decline within three decades coincided with the completion of the four lower Snake River dams, all constructed and operated by the Corps.

Scientific studies have consistently identified the development and operation of hydroelectric dams as the main cause for the decline of Columbia Basin salmon and steelhead. For example, NMFS concluded that the dams built since 1968 (three of the lower Snake dams and John Day) reduced adult spring/summer chinook smolt-to-adult returns from a sustainable four percent in the 1960s to less than one and one-half percent between 1970 and 1984.49 And the latest scientific evidence shows smolt-to-adult return rates to be less than one-half percent, at least four times below survival levels. The economic consequences of salmon declines have been devastating. Regional economic losses caused by salmon declines in the Columbia have been conservatively estimated at approximately $500 million per year and 25,000 family wage jobs.

While it is clear that the mainstem dams are the principal culprit in the decline of the Snake River salmon, there are divergent opinions on how to address the problem. For roughly the last twenty years, the Corps has conducted a smolt transportation program, under which juvenile salmon (smolts) have been collected at mainstem dams, loaded on trucks or barges, and transported down river for release below the dams.

Despite continuing declines in run sizes, NMFS has maintained this trucking and barging "experiment" under ESA procedures, promising to evaluate the program and decide whether to continue it in 1999. Since everyone agrees that the status quo is unacceptable, the 1999 decision basically involves a choice between two alternatives: 1) continuation of transportation with improved barging techniques, structural improvements at the dams, and likely increased flow augmentation; or 2) breaching the four lower Snake River dams, and also either drawing down or breaching John Day. The first option, improving the transportation program, superficially sounds appealing and is the most politically palatable, because it would minimize disruption of hydroelectric power generation and navigation. It has a glaring flaw, however: the Corps has employed transportation for two decades, but has failed to recover the fish runs. NMFS's scientists concede that the existing transportation will not reverse the decline of the fish runs, let alone recover them. An "improved" transportation program would rely on the development and installation of an expensive new and unproven technology, surface collectors, designed to increase the number of juvenile migrants collected at the dams, as well as reduce stress to the fish resulting from the collection process.

Even if this new technology works, it would not address stress associated with transportation itself, the genetic selectivity of the transportation program, or transportation's potential disruption of the smolts' homing capabilities. Further, surface collector test results are not encouraging and, even if problems can be overcome, development of the technology could take ten years, with an even longer period for testing and implementation Moreover, recent survival figures for transported fish are less than one-half of one percent adult returns from smolts in 1997 and 1998, far below the two percent thought necessary to recover the salmon runs. An "improved" transportation program is hardly likely t0 result in the 400 percent increase in survival needed to produce sustainable fish runs.

An attempt to restore salmon through an improved smolt transportation program will almost surely include increased flow augmentation to speed smolt travel to the collection facilities, as well as to improve the survival of fish that escape collection. Under the ESA, the Corps contributes about one million acre-feet of water from Dworshak on Idaho's Clearwater River, and the Bureau of Reclamation contributes 427,000 acre-feet of water from upper Snake River reservoirs.

These storage releases to supply fish flows remain quite controversial among Idahoans, prompting both judicial challenges and legislative action Moreover, since 1994, the Northwest Power Planning Council (Council) has called for an additional one million acre-feet of water to augment fish flows from upper Snake Basin reservoirs, A consultant's report indicated that this amount of water was available through a variety of sources, but the Bureau of Reclamation and NMFS have thus far declined to seek it. Although the Idaho Department of Fish and Game (IDFG) has recognized that increased flows is the only element of improved transportation program that has scientific data supporting it, the state of Idaho continues to oppose increased flow augmentation as ineffective and disruptive to other water users.

Nevertheless, the latest data shows a strong relationship between higher river flows, colder water temperatures, and salmon survival in the lower Snake River, particularly for juvenile fall chinook that migrate downstream in the summer. The other alternative—breaching the dams to allow natural river flows—has growing scientific support, although it faces considerable political skepticism. This natural river flow alternative has been called by the Corps's consultants "the biological option of choice if salmon and ecosystem restoration is the primary goal." The Council's independent group of scientists endorsed permanent reservoir drawdowns to expose and revitalize alluvial reaches currently drowned behind mainstem reservoirs in order to re-establish historic spawning grounds.

IDFG concluded that the "natural river option is the best biological choice for recovering salmon and steelhead in Idaho... with the highest certainty of success and lowest risk of failure, and is consistent with the preponderance of scientific data." Whether NMFS will choose the dam breaching option or the improved transportation option will not be known until at least December 1999. However, given the growing biological support for restoring natural river flows by breaching the dams, the key determinant to implementation of permanent reservoir drawdowns seems likely to be their perceived economic costs.

The Emerging Science of Restoring Natural River Flows by Dam Breaching

In recent years several scientific reports have called into question the efficacy of continued reliance on truck and barge transportation of juvenile salmon as the linchpin of Snake River salmon recovery efforts. These reports have suggested that recovery efforts ought to focus on restoring natural river conditions. Perhaps the most prominent of these studies is the peer-reviewed studies by the scientists participating in the Plan for Analysing and Testing Hypotheses (PATH), which endorsed the natural river option in 1998.

The Detailed Fishery Operating Plan, 1993

In November 1993, federal and state fishery agencies and treaty Indian tribes, through the Columbia Basin Fish and Wildlife Authority, prepared a comprehensive plan for operating the Columbia Basin dams in a manner compatible with salmon migration. The plan advocated seasonal reservoir drawdowns and flow augmentation to help restore more natural river conditions.

The fishery agencies concluded that barge and truck transport of juvenile salmon could not overcome poor river conditions created by the operation of the federal Columbia River Power System and noted that transportation had failed to halt the decline of the Snake River species.

The state agencies recommended restricting use of transportation to a "last resort"—to be used only after implementation of all other passage measures failed to provide safe passage for juvenile salmon. The tribes opposed continuation of transportation under any circumstances. The plan called for drawdowns of the lower Snake River reservoirs to "minimum operating pools" between April 15 and December 1 of each year, seasonal drawdowns that have subsequently been discredited as being of questionable efficacy given their cost.

The plan also endorsed significant amounts of flow augmentation both from Dworshak Reservoir on the Clearwater River and from reservoirs in the upper Snake Basin, calling for 927,000 acre-feet from upper Snake reservoirs in 1994, and 1.927 million acre-feet in 1998. A 1994 report of the Council's Snake River Basin Water Committee indicated that over a million acre-feet of water, above and beyond the 427,000 acre-feet for flow enhancement called for in the Council's 1991 program and NMFS's 1993 Biological Opinion (BiOp), was available through a combination of initiatives, including a conjunctive ground and surface water program, dry year water leases, and land fallowing.

The Independent Peer Review of Transportation, 1994

After NMFS's 1993 biological opinion on the operation of the federal Columbia River Power System relied heavily on truck and barge transportation—an opinion that caused a federal district judge to suggest that salmon recovery efforts needed a complete restructuring—representatives of NMFS, the U.S. Fish and Wildlife Service, state fisheries agencies, and treaty Indian tribes convened a peer review panel to study the scientific issues of the transportation program. The ensuing report, completed in May 1994, concluded that the transportation program approved by the NMFS BiOp was "unlikely to halt or prevent continued decline and extirpation of listed species of salmon in the Snake River Basin."

The report noted that while transportation provides "a temporary respite" from dam-related mortalities, it fails to protect salmon from mortalities associated with river system conditions that exist throughout the salmon life cycle. Therefore, the report explained, "[i]n terms of effecting a program of salmon recovery, it really does not matter if the transported salmon survive at a higher rate than untransported salmon, unless the overall survival rate for the population is sufficient to [recover the species]."

The Independent Peer Review criticized the transportation program for proceeding in the absence of a standard for hydroelectric project passage survival, observing that this lack of a standard made the utility of transportation "highly speculative," despite preliminary indications that some species of salmon seemed to benefit from transportation under the river conditions existing at the time the transportation experiments were conducted.

Moreover, the report questioned a key assumption of the preliminary studies showing that transportation benefits salmon explaining that "[f]ish appeared to be similarly handled and marked" regardless of whether they were transported or designated for in-river migration. Thus, the report suggested that the key premise underlying studies showing that transportation could benefit salmon—that fish handled and marked and returned to the river as in-river "control" fish suffered no mortality as a consequence of this handling and marking—had not been adequately tested and could be flawed.

This fact, coupled with transportation's potential to induce aberrant homing behaviour, adverse genetic effects, and stress, resulting in increased vulnerability to predation and disease, led the scientists to conclude that "[a]vailable evidence is not sufficient to identify transportation as either a primary or supporting method of choice for salmon recovery in the Snake River Basin."

The 1994 Amendments to the Columbia Basin Fish and Wildlife Program

The 1994 amendments to the Columbia Basin Fish and Wildlife Program adopted by the Northwest Power Planning Council (Council) incorporated a phased approach to Idaho Governor Cecil Andrus's suggested seasonal drawdown plan. The amendments also called for undertaking "a mainstem experiment" in which an approximately equal number of fish would be transported as would be allowed to migrate in-river. This experiment would require decreasing the number of fish transported to allow a comparison of the ability of transported versus in-river fish to produce returning adults. Except for the mainstem experiment, the Council voted to restrict transportation to "extremely adverse" river conditions, essentially low water years. Although the Council did approve transportation in the short term, especially under low river flow conditions, it warned that barging and trucking salmon "should not be regarded as a substitute for changes in the river ecosystem" or "as a device to delay substantial improvements in in-river survival conditions."

The National Research Council Report, 1995

In 1992, the National Research Council formed the Committee on Protection and Management of Anadromous Salmon to review the population status, habitat, and environmental requirements of Pacific salmon species in the Pacific Northwest. The committee gave only qualified endorsement to continued transportation in the short run, citing studies showing increased survival of transported fish compared to fish migrating in-river However, the committee failed to consider the fact, as recognized by the Independent Peer Review of Transportation, that in-river fish suffer stress as a result of marking and handling due to the transportation program, and it also seemed to assume that river conditions could never be improved. Nevertheless, the report did endorse a policy of reliance on natural river functions in the long run, endorsing a pragmatic approach to improving the situation that relies on natural regenerative processes in the long term and the selected use of technology and human effort in the short term... rather than on a primary emphasis on substitution, i.e., the use of technologies and energy inputs, such as hatcheries, artificial transportation, and modification of stream channels.

The Tribal Restoration Plan, 1995

In June 1995, the treaty fishing tribes of the Columbia Basin released Wy-Kan-Ush-Mi Wa-Kish-Wit (Spirit of the Salmon), their salmon restoration plan. The tribal plan completely eschewed artificial transportation of juvenile salmon, advocating instead permanent reservoir drawdowns to restore natural river functions. The tribes called for a long-term goal of achieving "mean historical flows" during the juvenile migration season and reducing dally and hourly fluctuations in flows occasioned by peak power operations. To help achieve these river flows and to restore ecosystem functions, the tribal plan called for permanent reservoir drawdowns.

The tribes recommended the immediate termination of artificial transportation, noting that halting transportation would allow the testing of alternative passage measures foreclosed because of the transportation program. The tribal call for the termination of the transportation program was hardly radical; a year earlier the four-state Northwest Power Planning Council also called for terminating artificial transportation, albeit only in the long term.

The Independent Scientific Group Report, 1996

The Northwest Power Planning Council established the Independent Science Group (ISG) in 1994. Comprised of eminent scientists and structured to insulate them from political influences, ISG's purpose was to analyse the science underlying the Council's fish and wildlife program and suggest a scientific conceptual foundation for the program. In 1996, ISG issued its report, which criticized the Council's program for lacking a coherent conceptual foundation, and recommended a program grounded on what it called "normative river conditions" or the restoration of ecological processes consistent with the needs of native fish and wildlife species. The scientists faulted salmon restoration efforts for relying on failed technological fixes, like hatcheries and artificial transport of juvenile salmon, and suggested that it was unlikely that technology would ever substitute for a natural river system.

ISG noted that these technologies were adopted with little or no scientific study and recommended that in the future such measures should bear the burden of proof, being implemented only after intensive evaluation. The report concluded that "available evidence is not sufficient to identify transportation as either a primary or supporting method of choice for salmon recovery in the Snake River Basin." This conclusion was the result of the scientists' finding that artificial transport cannot provide "the minimum survival rates necessary for the maintenance of population levels let alone those survival rates necessary for rebuilding of salmon populations."

Instead of barging juvenile salmon around the dams, the ISG report called for restoration of river flows as close as possible to the hydrograph that existed in the predam era and for maintenance and restoration of mainstem spawning habitat like the undammed Hanford Reach on the mid-Columbia, the last free-flowing stretch of the mainstem Columbia or lower Snake Rivers. To accomplish mainstem habitat restoration, ISG recommended permanent reservoir drawdowns, specifically suggesting that the John Day or McNary Reservoirs be lowered to expose alluvial reaches that historically supported salmon spawning.

The report also noted that drawdowns of the lower Snake reservoirs would be consistent with its normative river concept. The scientists called attention to the restoration of mainstem habitat, rather than tributary habitat, because it was the mainstem that historically supported metapopulations of salmon, and the scientists argued that restoration efforts should focus there. Restoring mainstem spawning habitat will require permanent reservoir drawdowns.

The Idaho Department of Fish and Game Report, 1998

In May 1998, the Idaho Department of Fish and Game (IDFG) issued a report on the causes of the decline of Idaho salmon and the available options for recovery. The report concluded that the primary cause of the decline of Idaho salmon was the construction and operation of dams built in the 1960s and 1970s. The report also observed that the transportation program "has not compensated for the dams and is unlikely to provide recovery." IDFG called for the establishment of a two to six percent smolt-to-adult survival standard for recovery and embraced the independent scientists' concept of a "normative river" as the best means to achieve this standard. The report observed that "available data provide no indication [the current transportation program] can sustain a 2-6% smolt-to-adult survival" and noted that current operations also fail to meet both 24-and 100-year survival standards.

With status quo not a viable option, the IDFG report considered two available alternatives: an enhanced transportation program or restored natural river flows. It rejected transportation for three reasons: 1) uncertainties over the efficacy of untested surface collectors (which would be at the heart of the program), 2) the fact that it would take over a decade for surface collectors to be tested and installed, and 3) the fact that an enhanced transportation program would doubtless require additional Idaho storage water to flush salmon smolt through the reservoirs to collection stations. IDFG concluded that existing "data does not indicate flow augmentation can provide enough survival benefits for recovery," citing Idaho's comments to NMFS claiming that " istorical water velocities cannot be attained with the current reservoirs, even using all reservoir storage in the basin."

As a result, IDFG endorsed the natural river option, noting that it had "a high likelihood of meeting recovery standards" if in fact barging has not compensated for the effects of the dams, which IDFG determined it had not. The report concluded that "the natural river option is the best biological choice for recovering salmon and steelhead in Idaho. This assessment is logical, biologically sound, has the highest certainty of success and lowest risk of failure, and is consistent with the preponderance of scientific data."

The PATH Preliminary Decision Analysis Report and Weight of Evidence Workshop, 1998

The Plan for Analysing and Testing Hypotheses (PATH) is an interagency working group of twenty-five scientists created by NMFS's 1995 BiOp. The BiOp charged PATH with evaluating alternative models for Snake River salmon recovery in order to help NMFS make its 1999 decision on continued transportation versus dam breaching on the basis of the best scientific advice possible. PATH is to review the scientific uncertainties affecting salmon survival and "using expert judgment, based on all existing evidence, to quantify the relative degree of belief in... conflicting hypothesis about the effects of management actions on stock performances." PATH studies follow structured scientific procedures developed by scientific consensus and are peer reviewed. In a March 1998 report, PATH scientists compared three alternatives under three computer models: 1) status quo, 2) maximum transportation without surface collectors, and 3) natural river drawdown. Under one computer model, Fish Leaving Under Several Hypotheses (FLUSH), the natural river drawdown was the best alternative, but the other model, Columbia River Salmon Passage (CRISP), found drawdowns to be the worst alternative.

A revised analysis in August 1998, however, concluded natural river drawdown was the best alternative for recovery in the long-run, with close to one hundred percent likelihood of recovery over forty-eight-year and one hundred-year time periods. The CRISP model remained considerably more optimistic about short-term (twenty-four-year) recovery under the transportation alternative than the FLUSH model (sixty-one percent chance of meeting the survival standard versus a ten percent chance), but it still found natural river drawdowns to be superior to transportation (seventy-six percent chance of meeting the survival standard versus a sixty-one percent chance). The FLUSH model produced a much lower probability of survival under the transportation alternative, with just a ten percent chance of meeting the survival standard in the short run (twenty-four-year period), a twelve percent chance in the mid-term (forty-eight-year period), and a thirty-seven percent chance in the long run (one hundred-year period). After analysing the August 1998 PATH results, staff of IDFG determined that, in light of recent smolt-to-adult returns of transported fish of less than one-half percent, survival rates remain four to twelve times below that required for salmon recovery.

4

Managing Water Diversion

Water is increasingly a major political issue as scarcity of the resource grips several societies, especially in developing countries where agriculture is responsible for about 80 percent of water consumptive use. However, western nations are not immune to water tensions, for instance in Greece, Spain, Italy, or the western United States, where the available water is being exploited to the limit: the Colorado no longer reaches the sea, and growing debates are emerging as to whether water should be allocated to thirsty cities or to agriculture; whether public funds should be invested again to increase the resource or demand management implemented; and whether water could be imported from far away.

These questions are increasingly relevant: consumption patterns of water in the western US are clearly not sustainable. Given the technology available today, massive water transfers could only come from Canada. There have been several mooted projects in this vein, mostly in the 1960s and 1970s. Should Canadians worry about water exports to the United Sates, especially in the frame of NAFTA? Or is climate change going to be the decisive factor in the debate?

The Advent of Water Addiction

Water is a key ingredient in the fabric of the western American society, as has been well studied by Donald Worster. The west is not completely waterscarce, for several mighty rivers flow in the region, mainly fed by snow and glaciers from the Rockies; but it definitely is a semi-arid region, compelling all societies living there before the industrial revolution to adapt to water scarcity. Early 20th century American society, empowered by the industrial age, decided to harness rivers and aquifers. Technology enabled American society to eliminate the water scarcity burden and developed the illusion that technology would always bring about a solution to growing water needs. "What nature does not yield freely, humanity should refashion to better suit human needs.... Nature has no greater purpose than to serve Man, and Man has no greater purpose than to work the land and take his place in the productive cycle."

Wendy Nelson Espeland also clearly depicted the representation that developed at the time that all water flowing unused to the ocean was a wasted resource, and that rivers needed to be "tamed" and "harnessed" to be put to use. President Franklin Roosevelt declared in 1935, when inaugurating Boulder Dam, that "the mighty waters of the Colorado were running unused to the sea. Today we translate them into a great national possession." This idea is still very much alive today. "Much of the contemporary culture of the American Southwest is based on denying its desertness.... In a subdivision being built to the south [of Las Vegas], Paseo Verde Parkway and Val Verde Road intersect in Green Valley Ranch. The concept of green, like sod lawns, was an imported fantasy."

Technology can make up for water scarcity and water must be used so as to develop western resources: "The rural Western ethic is that all wealth comes out of the ground, either as grass growing or as minerals being mined.... So the fact that today's reclamation projects-such as Garrison, CUP, Animas-La Plata-cost a few million dollars for each farmer they put on the land, doesn't cause their proponents to blink. That, they say, is the price society pays for creating the stuff of wealth. Without it and the other industries based on earth, there is nothing."

Early Massive Water Transfer Projects

Massive water transfers were first built in the eastern United States in 1847 with the Croton aqueduct for New York City and the 1900 Chicago diversion. As early as 1906, water transfers began being built in the west, with a 46 km-long canal from the St. Mary River in Montana to the Milk River, triggering a severe dispute with Canada that forced the United States to negotiate the 1909 boundary waters treaty. Other projects soon followed: the Los Angeles aqueduct in 1913, the Hetch Hetchy water supply (1934), the Colorado River aqueduct (1941), the Ail-American Canal (1942), and the California central valley project (1951). Diversions were initially designed for municipal water supply, but irrigation soon became their main rationale, as the sector benefited from generous subsidies from the federal government through the Bureau of Reclamation (created in 1902), its agency dedicated to the development of agriculture in the west.

When exploited resources began to show signs of exhaustion, engineering firms started to design huge transfer projects from Canada. But the main trigger of the idea that water resources were exhaustible and not adequate to sustain limitless development in the western United States struck in 1963 when the supreme court compelled California to stop excessive Colorado water pumping.

Colorado River water had been apportioned between upper-basin and lower-basin states with the Colorado compact of 1922, but Arizona felt California tried to prevent its own use of the water. With the 1963 ruling, California's share was established at 4,4 million acrefeet (af), by far the largest share in the 15 million af available, but a figure that set a limit to what California could bank on. The very idea that the water resource was all of a sudden set at a strict limit sent shockwaves throughout California that triggered a frenzy of plans designed to increase supply through continental water diversion projects from Canada.

These projects, like the earlier water diversion schemes, involved diverting large volumes of water over long distances, but their sheer size placed them on a very different scale, both regarding their length (often greater than several thousand km) and their cost (several dozens of billions of dollars). They were designed, from 1959 to 1985, in the United States and sometimes in Canada, on a magnitude no project had ever reached before. These projects involved the transfers of the Mackenzie River, for instance, all the way through the flooding of the Rockies Valley to the Mississippi and Colorado Rivers, or the diversion of the Great Lakes, or even the damming of James Bay so as to convert it into a freshwater lake and its subsequent transfer to the American west. Costs involved reached hundreds of billion dollars and environmental impacts were not studied at the time.

Massive Continental Water Diversion Projects Never Implemented

Although these continental water diversion projects received considerable media and state government attention in the west in the 19605, none went beyond the feasibility study level and none managed to get an official approval.

Cost

Water is heavy. When transported by aqueducts over long distances, it requires a lot of energy merely to be moved and pumped over obstacles. The Grand Canal, designed in 1959 by Canadian engineer Tom Kieras would have required at least 6 nuclear plants merely to pump the water from James Bay to Georgian Bay in Lake Huron. It is therefore expensive to operate these infrastructures. Moreover, the capital requirement is huge, as attested by their estimated construction costs at the time of the project design. Governments were more than reluctant to invest such large amounts of money for water exports, especially when the profitability of such projects was far from certain.

Estimations of the cost of water transported or produced by different means show that desalination is becoming much cheaper than it used to be: about $0.6 to $1.2 per cubic meter, as compared to $0.55 to $1.35 for plastic container transportation; $1.25 to $1.5 with water-carrying ships; $0.6 to $0.85 with iceberg transportation; $0.07 to $1.8 for water recycling; and $0.8 to $3 for transfer canals of about 500 km. It is true these means do not all serve the same purposes, as desalination plants are usually located by the sea and serve urban demand. Sea water desalting technology, in particular, improved so fast between 1985 and the present that operation costs have been reduced dramatically and divided by five, from $2.5 to $0.55 per cubic meter.

Desalting is now a very affordable water-producing technology for urban and industrial consumers, enabling them to tap into an inexhaustible source. However, the water it produces cannot be considered for irrigation purposes, given its cost, its distance from the interior, and the large volumes irrigation demands. But these figures are useful inasmuch as they show that water transfer by canals is not as inexpensive as many advocates insist it is: it is not certain that farmers could afford their expensive water volumes without massive subsidies.

Stabilizing Demand in the United States

The main reason that the giant diversion schemes never came to fruition is that demand was not really present in the United States for Canadian water. Several factors explain why these projects lost their allure for both governments and society. The budget crisis that began in the early 19705 in the United States precluded public money for such large endeavours, at a time when public opinion gradually turned against large scale engineering projects that tampered with the environment: 1977 was the "hit list" year when President Jimmy Carter discarded several water development projects in the west because of their extremely low benefit/cost ratio. The economics of continental diversion projects have so far worked against them and will do so for several more years.

It was all the more difficult to argue for these continental diversion schemes as water use, contrary to what had been expected in the 1960s, showed a definite trend toward stabilization, especially in agriculture. More cost-recovery-oriented water pricing; competition from other regions, mainly Asia; and cost incentives that lure American producers to Mexico are among the factors that explain why water use in agriculture remained roughly stable between 1975 and 2000 throughout the country.

If the federal government does agree, during the Doha round of trade negotiations, to reduce agricultural subsidies, water prices for farmers could increase markedly, resulting in financial incentives to consume less. Alternatively, water demand for irrigation could decrease as the result of farmers going out of business. As a whole, water withdrawals in the US have increased slowly, much more slowly now than its population, and could even begin a downward trend should competition from foreign fruit and vegetable producers work against local farmers.

In addition, although there still is room for improvement, water use per person is showing signs of stabilization, probably thanks to both tariff and education policies. The evolution of water use in the United States, from 1970 to 2000, shows that although the population grew by 38.6 percent, water use increased by only 10.7 percent. Urban consumers began to value water conservation with the development of urban water use reduction programs in most large cities. Industrial use actually decreased by 55 percent, whereas irrigation grew by only 6.1 percent.

Therefore, although water is still used at an unsustainable rate in the western United States, importing water from Canada is not as urgent as appeared to be to some politicians a few years ago, such as former senator Paul Simon. A stabilizing trend in water withdrawals and the availability of cheaper sources with desalting plants led public planners to forget about massive water transfers from Canada. When asked, even the Western States Governors Association estimated continental diversion projects were unlikely to ever be built."

Opposition from Within the United States

Moreover, there also was opposition from within. Governmental archives from the early 1980s attest to the western United States lobbying for the diversion of Great Lakes water to quench its thirst. The International Joint Commission, created to prevent and resolve disputes between the United States and Canada under the 1909 boundary waters treaty, explicitly warned against water diversions from the Great Lakes basin in its 2000 "Final report on protection of the waters of the Great Lakes."

Great Lakes states wanted to resist these projects, both for environmental and political reasons: why would the Great Lakes states give California added value at a time when so many firms were leaving the area and moving to the West Coast? The Council of Great Lakes Governors (CGLG), created in 1983, is a partnership of the governors of the eight Great Lakes states and the two Canadian provinces of Ontario and Québec. The Great Lakes charter stemmed from their growing concern that Great Lakes water could be diverted to water-scarce regions of the United States.

The charter, signed in 1985 by the CGLG members, created a notice and consultation process for Great Lakes diversions. The signatories agreed that no Great Lakes state or province would proceed with any new or increased diversion or consumptive use of Great Lakes water over five million gallons per day without notifying, consulting, and seeking the consent of all affected Great Lakes states and provinces.

After an Ontario firm, Nova Group, tried to tap water in 1998 from the Great Lakes for export purposes, triggering an uproar of protests from both the Canadian and the American sides of the lakes, the CGLG decided to further strengthen the water export ban. This led to the Great Lakes charter annex, signed in June 2001.

The annex outlines a series of principles for reviewing water withdrawals from the Great Lakes basin that is grounded in protecting, conserving, restoring, and improving the Great Lakes ecosystem. In December 2005, the annex 2001 implementing agreement was signed, with specific guidelines for the enforcement of the principles. It will also form a legal basis for the ban when the Great Lakes-St. Lawrence River basin water resources compact is ratified by the US congress, making it legally binding for American Great Lakes states. With this agreement, water diversions are only permitted over short distances from the Great Lakes, must ensure that water is returned to the Great

Lakes basin, and are only permitted if there is no other solution to satisfy the water demand.

Massive Transfers

Canadians became very fearful in the late 19905 about potential American water transfer schemes, leading to calls for the federal government to enact a ban on water exports. Wisely, Ottawa refused to enact any law that would implicitly recognize water as a good, but rather put forward bill C-6, voted in December 2002, which forbids interbasin transfers. Under NAFTA, this was a clever move inasmuch as it leaves water outside the realm of commerce and applies to both Canada and the United States. But Canadians were ignoring the fact that large-scale water transfers already exist throughout the world, and especially in Canada. Several massive transfer schemes have already taken place: in British Columbia (the Kemano transfer, 1954); in Manitoba (the Churchill derivation, 1976); in Ontario (the Long Lake and Okogi River transfers, 1939 and 1943); in Labrador (the Churchill Falls transfers, 1971); and in Québec (the Caniapiscau and Eastmain transfers, 1985). There are major differences from American water diversion projects: Canadian transfer schemes were for hydroelectric development purposes, and not for urban consumption or irrigation. They are usually massive and over short distances, less than 200 km, whereas American water diversions often run for more than 500 km. There are currently plans to divert the Peace River in northern Alberta to meet growing agricultural and urban water needs, but that project is far from accepted. In addition, HydroQuebec is currently developing other water diversion projects for the Betsiamites, Eastmain, and Manouane rivers development.

There is controversy over the extent of the negative impact of these diversions on the environment, since the concept of minimum ecological flow remains controversial among biologists. Water does remain in Canada for each of these diversions, but this is not the point: Canadians collectively ignore the fact that their daily comfort and economic activity rest, to a certain extent, on major river diversions. It is therefore difficult to defend the argument that water transfers to the United States would weaken the environment, if one refuses to consider the phasing out of transfers within Canada.

At the end of 2005, with the signing of the annex 2001 implementing agreement, it was safe to say continental water diversions were definitely outdated as a water management concept. Demand was stabilizing in the western United States; public opinion had developed a strong distaste for the idea; governments were more than reluctant to invest the billions of dollars each scheme implied; diversion were by far not the most cost-effective way to provide for additional water volumes; and conservation and recycling were promising alternatives to the long-used supply management approach.

Toward The Return of Massive Continental Water Diversion Projects

Climate change could shake this status quo. As years go by, and as scientists as well as local governments accumulate impressive data attesting to a definite trend toward climate change, the developing weather shift could prove more challenging than the long-advocated water exports to the United States of the 19605 and 19705. The possibility that climate change could affect Canada's water quality and quantity is serious enough that Environment Canada is spending several million dollars on scientific studies to try and assess the risk. Competing uses-for urban areas, agriculture, and oil production, for instance-in regions such as Alberta where water could become scarcer, could lead to conflicts between groups that would see authorities intervene with unpopular measures such as water pricing or regulation.

Present forecasts, so far proven by observed data, show increased global precipitation, but this extra amount of water would be more than offset by increased evapotranspiration (Et) because of higher temperatures. In the already dry Okanagan Valley in British Columbia, water demand for orchards could therefore be multiplied by 12 in the summertime because of increased Et. In addition, a change in the pattern of how precipitation falls throughout the year, coupled with changes in temperature, could very much alter its availability, resulting in seasonal but dramatic supply reduction, as happened in Québec in August 2002.

In the Prairies, already less snow falls in the winter, and it melts earlier in the spring, allowing for less water in rivers in late spring and summer. Along the eastern slopes of the Canadian Rocky Mountains, glacier cover has decreased rapidly in recent years, and total cover is now approaching the lowest experienced in the past 10,000 years. Rivers in the Prairies are mainly fed in the summer by melting glaciers. As the glacial cover has decreased, so have the downstream flow volumes. This finding appears to contradict projections of the intergovernmental panel on climate change (IPCC) that warmer temperatures will cause glacial contributions to downstream flow regimes to increase in the short term. Historical stream flow data indicate that this increased flow phase has already passed, and that the basins have entered a potentially long-term trend of declining flows. The continuation of this trend would exacerbate the water shortages that are already apparent across many areas of Alberta and Saskatchewan owing to drought. Climate change is of course not merely affecting Canada, but the United States as well. Similar conclusions have already been reached with regard to the potential impact of the Rockies' disappearing snow and glacial cover on the western United States. Planning for the impact of climate change remains difficult, because building accurate models for various regions-ones that give a true picture of future available runoff-remains a daunting task for programmers. What we now have are educated guesses sustained by general models and data trends. But these trends do show that several regions in the United States, especially in the dry west, could be faced with critical water shortages, given present demand.

This might prove to be a scenario in which the United States could become interested again in Canadian water: the goal would be to prevent a social crisis triggered by conflicts between agriculture and cities for dwindling water resources. The equation for the American federal government would be the perceived financial, social, and political costs of finding new water resources, as opposed to arbitrating against agricultural water needs. No one can predict what choice officials would then make if this scenario came true. The following parameters would be significant:

- Agriculture is responsible for 80 percent of water consumption in the American west.
- A 15 percent reduction in irrigation use would satisfy the needs forecast for cities and industries."
- Agriculture does not currently pay the market price for its water, thanks to strong federal subsidies; tariff increases would result in more disciplined behaviour and encourage conservation.
- Western farmers have begun selling or transferring their water rights to cities, as illustrated by the October 2003 water deal between the Imperial Irrigation District and the Metropolitan Water District in California.

Massive water diversions are not new in North America: they were extensively used in both Canada and the United States, in the latter mainly to develop the west and to bring water to cities and fields so as to sustain an imported way of life, and in the former to develop hydropower projects. Beginning in the 19605, western American states and engineering firms began developing continental water diversion projects because of fears that continued development would bring about severe water scarcity and social disruption. These doomsday scenarios did not materialize, due to reduced economic growthespecially in agriculture, as well as more efficient water use technology and the advent of water conservation, and the huge environmental and financial costs of these projects.

However, the very idea of massive water diversions is not completely dead. Tom Kierans, the designer of the Grand Canal, is still lobbying Ottawa and Washington. More challenging is the idea that ongoing climate change could dramatically affect agriculture in the American and Canadian west. Under these conditions, what answer would governments have? Whither agriculture? Would governments favour the more profitable and water-efficient farmers or bring back to the forefront the idea of transferring large amounts of water from humid parts of the continent to drier zones? The answer probably lies in the severity of the developing weather pattern change and in the adaptation capability of the agriculture sector.

Domestic Water Supply in Rural Areas: Implications for Policy

Lack of access to safe water is at the heart of the poverty trap, especially for women and children, who suffer in terms of illness, drudgery in collection of water, and lost opportunities because of the time that water collection consumes. In rural Africa, according to the World Bank, 40 million hours are spent, each year in collecting water for domestic use and half of Africa's population is without access to safe water (Black, 1998). Recent renewed focus on poverty alleviation has resulted in increased attention to the benefits of improved water accessibility. Poverty assessment research has consistently shown that improvement in water services is a critical element in designing and implementing effective strategies for poverty alleviation. In Tanzania, experience with respect to rural water improvement has, in general, been an unmitigated disaster.

Water provision facilities have invariably fallen into disuse or disrepair because the approach used to run them failed to ensure sustainability of services. As a result, a new water sector vision has emerged based on the demand responsive approach (DRA). The major principle of the vision regards water as an economic and social good to be managed at the lowest appropriate level. The DRA recognises the inherent capacity of communities in taking greater responsibility for identifying and solving their water supply problems.

Adoption of the DRA by most water and sanitation programmes has changed the approach to evaluating their success and failure. Traditionally, water and sanitation programmes have depended on detailed blueprints to provide the basis for control and predictability. Benefit-cost analysis has been a major economic tool in evaluating these projects. However, the DRA broadens the scope of evaluation and blueprints cannot be drawn up since decisions are made jointly with communities, and problem solving is based on those partnerships (Narayan, 1995). Participatory evaluation becomes an essential tool in assisting attaining stated objectives and is an integral part of overall programme monitoring and evaluation activities. Participatory evaluation requires the utilisation of different tools and approaches in evaluating the sustainability of such projects.

This chapter is part of a follow-up participatory evaluation exercise of two community water supply programmes in Central Tanzania and involves presentation of the results on communities' willingness to pay (WTP) for improving the availability and quality of water services. WaterAid, a non-governmental organisation (NGO) based in the United Kingdom and the Lutheran World Federation through the Tanganyika Christian Refugee Service (TCRS), provide substantial support and expertise for such programmes. The former operates in the Dodoma Region while the latter operates in the Singida Region. The two regions constitute the semi-arid plateau of Central Tanzania.

It is anticipated that positive WTP indicates potential community ability to recover operation and management costs (Altaf et al., 1992). Thus WTP can be used to help ascertain the potential for fulfilling sustainability, at least from a financial viewpoint. A major objective of the study was to find out if communities studied were willing to pay an increased user fee or tariff for improving and expanding existing water services. If willing, how much (i.e. what percentage) could be added to the current user fee or tariff to achieve the required improvement? Based on the random utility hypothesis, a system of Multinomial Logit functions were developed and used to estimate individuals' willingness to pay and impact on revenue from increasing the water user fees or water tariff. The WTP was calculated by estimating the payments that would cause the respondent to be indifferent after a single unit increase in an independent variable. The estimated mean WTP and proportion of individuals who are able to pay were used to calculate the potential revenue that could be generated if the policy of increasing user fee or tariff was instituted. In Tanzania, only 40% of the population has access to clean and safe water (MoW, 1997). Between 1965 and 1985, the Government of Tanzania (GoT) spent a substantial amount of money in developing rural-based water prefects to alleviate the water supply problem, but without notable success. Most of the investments ended in failure. Government, in collaboration with NGOs, is currently trying to revive many of these projects.

The government is providing technical assistance, while the NGOs supply initial capital requirements, overhead costs and supervision in their design and implementation. When the utility is operational, the beneficiaries (i.e. the villagers) either pay for the water from the utility or contribute an annual water fee, both of which are added to the village water fund. It is anticipated that the size of the village water fund will be sufficient to operate and expand the water utilities, based on market forces. To attain such sustainability-related goals, existence of efficient water markets is essential. Sufficient funds have to be generated to meet both overhead costs and investment requirements. Information on WTP for required improvements provides an indicator of the communities' potential capability of generating funds for such purposes. Tanzania has sufficient surface and ground water to meet its present needs. Lack of capital and uneven availability of water across regions has limited Tanzania's capacity to utilise existing water resources to provide her population with clean and safe water. As population increases, conflicts in the allocation of water between energy, households, commercial livestock, and use in the agricultural sub-sector, increases. Weak water policies, inadequate community participation, and uncoordinated donor support has resulted in wastage or inefficiencies in the form of failed water development programmes.

After independence in 1960, donors supported large-and small-scale water development schemes. However, communities participated little in their planning and management, and also contributed little to both capital and operational/management costs. The policy of the government at independence was 'free, clean and safe water for all,' the objective being to provide clean and safe water to all villages in rural Tanzania by the year 2000 (MoW, 1997). As indicated above, this policy has now been abandoned in favour of a demand-driven water development programme. The new water policy enacted in 1997 (MoW, 1997), promotes the provision of efficient, affordable and sustainable water supplies and sanitation.

The emphasis is on community planning and management, private provision of goods, works and services relating to water supply, and public sector regulation, facilitation and environmental management. Water supply or water development programmes have to be demand-driven, and the community must be willing to participate in decentralised management and cost sharing. The Ministry of Water (MoW) has been restructured and its role has evolved from that of a water service provider to that of a regulator and a facilitator.

The new water policy underscores the importance of community participation and management in ongoing and new projects. In order to improve water supply services at the village level, the roles of MoW and other stakeholders can be summarised as follows (MoW, 1997):

- Village level. Small-scale water supply projects are to he operated and managed at the village level. Operation and management costs are to be met with funds raised within communities. The emphasis is on management by participation through formulation of village water committees that oversee and manage the utilities on behalf of community members. The communities agree upon the operational modes, with specific emphasis being paid to women's participation at all stages of water project development and management.
- District level. The water department at the district level is required to facilitate training of water managers and attendants/mechanics at the village level. The department is expected to maintain a pool of experienced technicians who will collaborate with village level mechanics in servicing and repairing established utilities. Another role of the water department at the district level is to make sure that necessary spare parts are available when needed.
- Regional level. The water department at the regional level has jurisdiction in providing guidance and making sure that government water policy and rules are adhered to. Apart from managing water supplies at the regional headquarter level, the department has to provide consulting services to districts and villages in terms of training, and facilitating availability and distribution of spare parts.
- National level. At the national level, the MoW has responsibility for financing and managing large-scale water supply programmes (especially in large cities), to train water professionals, and to finance maintenance units at the district

and regional levels. Other responsibilities are to standardise capital equipment and tools used in water development, to help facilitate availability of spare parts, and to coordinate donor and NGOs activities to make sure they follow Tanzanian water policies and local government rules and regulations.

The general objective of the water policy is to propagate the new vision of community participation and management based on the DRA. Communities arc expected to appreciate that water supply is no longer a free service provided by government, but rather government or other development agencies are available to facilitate and complement community efforts at meeting their own water supply needs. Thus, this vision is currently guiding the development of community-based water supply projects at the village level in the Dodoma and Singida Regions of Central Tanzania. Detailed descriptions and the rules pertaining to the programmes in each region can be found in Kaliba (2002) or the WaterAid website.

In both regions, financing for the water projects is thus divided among the beneficiaries, the government and donors. Villagers make cash contributions towards capital costs. They contribute time and labour, local materials and contribute to hospitality costs of visiting government staff. The initial amount of money contributed to a project by a village has been standardised and agreed by the major donors working in the regions.

The cash contribution is not currently expendable but acts as a reserve fund or safety valve for the project. The community-based water fund has to be increased using profit margins generated from user fees (Dodoma Region) or tariffs (Singida Region) when the project is operational. The government provides professional staff, an annual cash contribution towards the programme, and provides some transport and most of the construction equipment. Donor (WaterAid in the Dodoma Region and TCRS in the Singida Region) funding pays for the purchase of locally procured materials, government staff fieldwork allowances, training courses and running costs. The donors also procure all imported materials, equipment and vehicles, and employ technical and management back-up staff. In the Dodoma Region the village water utilities have emphasised deep boreholes and engine-driven pumps, and charge a tariff (i.e. a specific amount per litre of water used). In contrast, in the Singida Region, much greater emphasis is placed on shallow wells (8-10 m deep) and a user fee (i.e. a specific amount per year per household), is charged. In the Dodoma Region, between 1991 and 1999, 857 projects were implemented under the

WAMMA1 project funded by WaterAid. During this study period, 29% of the projects were not functioning. For those not working, 75% were waiting to be repaired, and 25% had permanent problems (e.g. missing engines, damaged pumps or dried up waterholes).

Useful indicators of progress are the growth in the populations served and village water funds voluntarily contributed by communities. The total population served increased from 0.27 million 1994 to 1.23 million in 1999. This is an increase of about 320% over a period of six years. During the same period, the contributions to the village water funds increased, in real Tsh terms, from about 0.24 million in 1994 to about 14.87 million in 1999. Since there is no evidence of a regional increase in purchasing power, and given the low income per capita in the region (BoT, 2000), this is a good indicator of the success of community-based water development projects in the region.

During that period, TCRS constructed 49 boreholes, 465 shallow wells, and three windmills. The number of villages with water committees increased from three in 1986 to 244 in 1995. In the same period, three villages did set up water committees without the help of TCRS. The number of villages with water funds increased from zero in 1986 to 229 in 1995. Although these numbers indicate the magnitude of time and money invested in water utility projects, they of course, do not provide by themselves any explicit evidence of the potential sustainability of the projects and/or progress towards achievement of sustainability goals. To elicit WTP information from individuals, the contingentvaluation (CV) method is used. There are three major procedures used in CV surveys, namely conjoint analysis, dichotomous choice and the payment scale approaches. The conjoint analysis procedure asks respondents to rate rather than to price alternatives. It allows for measurement of consumer preference between items with multiple attributes (Baidu-Forson et al., 1997). In the dichotomous choice approach, respondents are asked whether they would vote to change the provision of some public good at a cost of $X to themselves. The respondents answer yes or no. For the respondent who answers yes, the experiment is repeated by varying X, until the respondent says no. After that, the respondent is asked to state the value he/she is willing to pay for the good or service if provided. For the respondent, the amount mentioned is considered the maximum WTP for that good or service. The payment scale procedure allows respondents to choose a value or price (cost) on a given scale measure. A range of values are presented to the respondents and they are asked to identify one value or price (cost) they are willing to pay or incur to purchase the good or service being evaluated. The respondent's choice from the given scale indicates the WTP for the stated good or service. Positive values indicate positive demand, and zero values indicate zero demand for the good or service.

In all cases, the surplus benefits are calculated using the parameters estimated from a model based on the type of distribution attached to WTP. In most cases, the models and type of distribution to use are dictated by the type of questions imposed and responses obtained. Other good references are those of Rosenberger and Walsh (1997) for ordinary least squares models; Kenkel and Norris (1995) for models with log-normal distribution; Steven et al. (1997) for Tobit models; and a classical work by Haneman (1984) for Probit and Logit models.

CV surveys entail three operations: designing a questionnaire, conducting a survey, and analysing the results. The validity of the estimates depends on how skilfully these operations have been done (Christe and Schwab, 1995). Donaldson et al. (1997) suggest that the use of a payment scale is more valid than the open-ended or closed-ended approach when presenting hypothetical bids. The limitation of the payment scale is constructing scaling benchmarks. This limitation can be reduced through using available market and nonmarket information to construct the scales, as suggested by Donaldson et al. (1997). From the above review, it can be seen that there are different approaches to modelling WTP. Nevertheless, all procedures involve the respondent choosing one option from a range of other alternative services or goods based on their expectations. In order to develop models that are consistent with economic theory, the McFadden (1981) random utility hypothesis has become a standard approach to modelling WTP. As far as our study was concerned, the objective was to find out if communities are willing to pay an increased fee or tariff in order to get additional money for improvement or expansion of existing water utilities. If willing, how much (i.e. what percentage) should be added to the current user fee or tariff to achieve the required improvement? Field data were collected from 30 villages that participated in community water utility projects for a period of three years in Dodoma and Singida Regions (15 villages in each region) prior to the field surveys. Dodoma Region was divided into five clusters based on tribal distribution. From each cluster, three villages were randomly selected, differentiated according to accessibility. The results of past participatory rural appraisals (PRAs) conducted by WAMMA (some as early as the 1999) were reviewed and used in making adjustments to the questionnaire and in helping to develop pre-coded responses and socioeconomic indicators. In the Singida Region, due to transportation problems (part of the El-Nino legacy), only Singida Rural District was included. Rased on information gathered at the District and Regional Water Departments and the Regional Development Office, the district was divided into three agroecological zones. From each zone, five villages were identified to form a sample cluster. The Singida Regional Development Officer also had some PRA results that were used to fine-tune the questionnaire used in the study.

Two survey instruments were developed to gather the necessary information — the village checklist and the structured questionnaire. The village checklist solicited general information at the village level. This was done through meetings and group discussions attended by community members, government leaders, project managers, pump operators, extension agents, and water committee members. The major objective of the checklist was to get a general overview of the village projects based on group discussions. The discussion also helped to pre-test the structured questionnaire and in further adjusting the precoded responses. The questionnaire was administered to 225 individual respondents in each region. In each village, 15 respondents (at least five men and five women) were interviewed. The respondents were selected randomly from the village register and grouped based on age and social status of the individuals. Sample selection was stopped when there were enough sample respondents in each group. The two survey instruments and further information about the approach used in the study can be found in Kaliba (2002).

Regarding information on participation and management, several questions were asked. The questions related to the source of the project initiative, personal involvement in decisionmaking, labour and financial contributions, and the involvement of women in project design and implementation, emphasising participation in decision-making and informed choice. Questions were also asked regarding consumer satisfaction with the services and management of the water utility. The questions focused on: getting information on personal involvement in management; rating of the water managers' performance; knowledge of how the water management structure functioned; and the possibility of influencing change. To quantify respondent participation in project activities, and satisfaction in project performance, individual's responses were aggregated and used to calculate participation and satisfaction indexes as suggested by Sara and Katz (1998) and by the World Bank/UNDP (2000). Once again, further details on the approach used as far as this chapter is concerned can be found in Kaliba (2002).

On the WTP section, a respondent was asked if there was a need to improve the current water service. The answer was either 'yes' or 'no'. If yes, what kind of improvement was needed? The respondent was encouraged to identify only one improvement. After the respondent identified the type or choice of improvement required, she/he was asked to suggest an increase in tariff (for the Dodoma Region) or user fees (for the Singida Region) in order to cover the costs of the proposed improvement.

As explained above, the analysis was directed at identifying types of water services that need improvement and ascertaining the WTP for the improvement. In the Dodoma Region, the results concerning the water utility service can be summarized as follows: about 14% of respondents indicated that they were satisfied with the status quo (no improvement); and 81% suggested increasing pump capacity or buying a more powerful engine (thereby increasing the rate of water discharge). Thirty-three percent of respondents proposed increasing the number of water distribution points (water reticulation), and 22% proposed other improvements such as fencing, construction of new water tanks, construction of security houses, and construction of livestock watering points. The suggested tariff increase ranged from 5 to 100 Tsh/20 1. The existing tariffs averaged 20 Tsh, with a standard deviation of 9 Tsh (per 20 1 bucket of water) and on average amounted to a total of 895 Tsh/household/year.

In the Singida Region, 31% of the respondents were satisfied with the status quo, 80% wanted deeper wells that can produce more water, 29% wanted to increase the number of hand pumps in their village for water distribution purposes, and 10% indicated other types of improvement enumerated above. The suggested increase in a user fee per household per year ranged from 20 to 200 Tsh. The average user fee was 503 Tsh per year with a standard deviation of 170 Tsh. During the analysis, variables for Clusters 4 and 5 in the Dodoma Region and Cluster 3 in the Singida Region were not included in the models due to correlation problems. The likelihood ratio statistics that test for no effects was statistically significant at the 5% level in both the Dodoma and Singida Regions models. The positive and statistically significant variables in the Dodoma Region were family size and satisfaction in the performance of project activities. These results imply a respondent residing in a relatively large family or who was satisfied in the project's performance were more likely to answer 'yes' to the question regarding the need for water service improvement. A large family implies the need for more frequent water collection trips. Improvements relating to reducing congestion at watering points (increased pump capacity or number of watering points) are likely to reduce time and effort expended in water collection. Satisfaction with reference to project performance implied both an increased demand and willingness to commit resources for improvement.

Negative and statistically significant variables were age, wealth and cash contributions. Older people and richer respondents were thus more likely to choose to maintain the status quo. In the study area, older people are less likely to be directly involved in water collection activities, while rich respondents could conceivably have access to other water sources (e.g. have their own wells) or delegate others to collect water for them. The negative sign on the cash contribution variable can possibly be attributed to 'contribution fatigue.' Respondents who contributed more during project initiation or development were more likely to say 'no' to the improvement question.

Finally, respondents in Clusters 4 and 5 were more likely to be predisposed to commit resources to improvements, because the general performance of the projects in these clusters was relatively poor (Kaliba, 2002). The underlying implications of the above results for water utility projects in the Dodoma Region are that projects whose performance creates a sense of satisfaction among consumers and a strategic water fund that is rising (so as not to create contribution fatigue) are critically important ingredients in generating community-related resources for water-related improvements. In the Singida Region, females were more willing to pay for improvement than male respondents, not surprising since they are primarily responsible for water fetching activities. The fact that this variable was not statistically significant in the Dodoma Region was unexpected. However, in the Dodoma Region water utility projects produced a more reliable water supply and there were generally fewer complaints about their performance (Kaliba, 2002). Another important and positively significant variable was the level of satisfaction concerning the performance of the project. The Multinomial Logit model was used to model factors affecting choice of services: maintain the status quo, increase rates of water discharge (pump capacity or depth of water well or more powerful engines), increase the number of water distribution points/water reticulation, and other improvements, as explained above. In constructing the Multinomial Logit models, the parameters of status quo variables were set to zero. Again, the signs on the estimated parameters show the direction of the marginal effects on choice of improvement desired. The comparison is related to the deleted choice. As an example, for Dodoma Region, respondents who perceived that water service quality was high were more likely to say 'no' for an increase in water supply or the number of water distribution points but were more likely to say 'yes' for other improvements. The interaction between quality and proposed bids were important in making the choices or indicating the type of improvement desired.

The opposite was true for the Singida Region where the interaction between quality and bids influenced the choices negatively. There, the important factor that significantly influenced the choice of improvements was individual bids. This may be related to the technology and poorer level of services in the Singida Region. Respondents may not be willing to spend more money on failing projects. Singida Region projects will have to increase performance in terms of quantity and quality of water supplied, in order to motivate community members to contribute more resources toward project development and expansion. This was particularly important in some villages where community participation in project initiation development was low (Kaliba, 2002).

Again, the reference is the status quo bid. The value of-0.145 in the quality variable in Cluster 1 of the Dodoma Region implies an increase in the quality of water by one unit decreases the percentage of respondents who want increase in water supply by 14.5%. About 64% of respondents in Dodoma and 59% of respondents in Singida voted for an increase in water supply through increased pump capacity or number of water distribution points (water reticulation). Most villages have only one watering point and it is clear that a lot of time and effort is required for water-collection activities.

5

Participatory Irrigation Management

PIM: What and Why

A Process of Improving Productivity and Sustainability of Irrigation Systems

Participatory Irrigation Management (PIM) refers to the involvement of irrigation users in all aspects and all levels of irrigation management.

"All aspects" includes the initial planning and design of new irrigation projects or improvements, as well as the construction, supervision, and financing, decision rules, operation, maintenance, monitoring, and evaluation of the system.

What are the Problems of Irrigation

In a national seminar on PIM in India held in 1994, the participants identified the following problems, or issues, with irrigation management. While one can debate whether some of these are "problems" or "symptoms" of problems, they are certainly issues familiar to all of us:

- Inadequate water availability at the lowest outlets
- Poor condition/maintenance of the system
- Lack of measuring devices and control structures
- Inadequate allocation for O&M
- Inequitable distribution of water
- Lack of incentives for saving water
- Poor drainage.

In particular countries and particular irrigation systems, some problems may be more critical than others. There are always additional problems or issues that could be added.

How can Participation Improve Irrigation?

Now let us consider the problems with irrigation management and think about whether and how participatory management approaches can help solve them. We will consider them one by one.

- Inadequate water availability at the lowest outlets
- ? Poor condition/maintenance of the system
- Lack of measuring devices and control structures
- Inadequate allocation for O&M
- Inequitable distribution of water
- Lack of incentives for saving water
- Poor drainage.

This new perspective of irrigation management has been applied to both projects as well as policies and has greatly expanded our understanding of why some irrigation systems work better than others. It is not only a physical problem; irrigation also involves a much broader set of management issues. For example, how is the system financed? What type of monitoring is carried out and how is this fed back into the O&M process?

Indeed, if we do a straight line projection of what is happening, we would perhaps predict that the state will eventually disappear entirely from the irrigation over the next several decades. But no one is realistically predicting the demise of the state's role in managing irrigation; there will continue to be an essential management role for the state. What is happening is a rationalization of the respective roles of government and users. The pendulum that had shifted far towards the direction of strong government involvement in all aspects of irrigation management is now swinging the other way, towards a more sensible equilibrium.

Inadequate Water Availability at the Lowest Outlets

If the water availability is inadequate because there is inadequate supply, then there are clear limits within which to suggest improvements. But even here, participation in the sense of transparency and dissemination of information about the water situation (e.g., through a system-wide organization) can engender good will from farmers vis a vis the system managers, thus minimizing the likelihood of intentionally breaking gates to increase water supply.

When the users understand the constraints faced by management, they are far more likely to cooperate. If the water availability at the lowest outlets is inadequate because of water thefts upstream, then part of the solution could entail bringing upstream and downstream users into a single management entity that would be accountable to all the users. This principle of including as many stakeholders as possible in a single organization underlies the very large water user organizations found in Mexico, for example. Participation of the users in managing the system is not the only possible solution, of course. More diligent management by government irrigation engineers could also solve the problem. But what are the incentives to an government engineer who will be paid by his agency at the same (usually low) rate regardless of how diligently he supervises water use between the upstream and downstream users.

Compare this situation with that of an irrigation engineer hired by a water user organization to oversee water distribution between upstream and downstream users. If he does not satisfy both sets of farmers, he will lose his job, and, at least in the case of Mexico, his salary may be double that of his agency counterpart. This engineer has a very strong incentive to distribute water equitably throughout the system, so that he remains in good favour in the eyes of all farmers.

Poor Condition/Maintenance of the System

The almost universal problem of deferred maintenance resulting in deteriorating infrastructure is on everyone's list of irrigation management concerns. Can participation help? When the users of the system are also in charge of maintenance, they have every incentive to perform timely repairs, to monitor the quality of the work, and to protect structures from vandalism. When the users also participate in initial design decisions, with the understanding that they will inherit the system as their own, they have a strong incentive to demand a system whose operation poses no technical problems, to monitor construction quality, and if they will repay some of the capital costs, to minimize the overall costs of construction. Most often this translates into reduced demand for lining canals except where it makes good operational sense.

The conservative tendency on lining which participation encourages may in itself result in substantial cost savings initially, and long-term improvements in maintenance (since users are reluctant to repair lining that they do not consider essential).

Lack of Measuring Devices and Control Structures

When irrigation managers are engineers employed by the government irrigation agency, there is little incentive to measure actual water flows. The philosophy of "no news is good news" is the standard approach to measurement. Similarly, control structures require both operational attention and maintenance which may be beyond the interest and/or budgets of the managing engineer. Would this situation change if farmers were in charge of management?

Imagine a situation in which the managing engineer would have to prepare weekly reports to a governing board of farmers, to whom the engineer owes his job. Would he now have an incentive to know how much water is going into each outlet? Would he now have an incentive to control the water so that he can adhere to his operational plan? This type of performance-based management by which the governing board holds the engineer accountable for water flows is theoretically possible under a government administered system. But if we can imagine two governing boards of an irrigation system, the first comprised of elected farmers who themselves depend of that irrigation water, and the second comprised of civil servants who enjoy job security, which of these boards is more likely to demand high performance? Which of these boards would be more likely to be interested in the flow rates in various reaches of the system?

Inadequate Allocation for O&M

Inadequate allocation implies either high costs, or low recovery, or both. The costs of both operating and maintaing the system are increased when farmers deliberately intervene to disrupt the scheduled operation, or to damage structures, and costs are reduced when farmers respect the operating plan and report maintenance problems promptly. Cost recovery appears to be enhanced when farmers can see that the money is going into system O&M (see Vicious Cycle of Public Irrigation), and when they perceive that they are receiving valuable services from the system, i.e., the timely provision of water. Can participatory irrigation management contribute to lower costs and higher recovery? When farmers are directly involved in the design of the system itself and the placement of structures and when they have input into establishing the operational plan, they will be more likely to both understand and protect the facilities and the plan.

When farmers are clearly the owners of the physical system, so that the maintenance costs are their own responsibility, they will have a strong incentive to protect the physical integrity of the system to reduce their overall costs. The cost recovery issue is complex. When farmers are involved in management they have a better understanding of how the funds are being used, and this knowledge is likely to render them more willing to pay the fees. To the extent that irrigation management is improved with user involvement, the farmers may feel more willing to pay for improved service (although these links are difficult to prove). Perhaps the best evidence for a relationship between participatory management and cost recovery comes from actual experience. In the Philippines, the payment rate for irrigation service fees is significantly higher in those systems where water user associations have been formed. In Mexico, payment rates are close to 100% because the system managers — controlled by a board of irrigation users — will refuse to give water without prior payment. Political realities would not allow this type of ultimatum were the managers part of the Irrigation Department.

Inequitable Distribution of Water

Disparities in the supply of water available to farmers are not only due to the physical problems of reaching the tail end of large systems. Disparities may also be caused by intentional tampering with the system, either by physically enlarging outlets, or offering bribes to the water masters. The result is that influential farmers can take more than their due share, thus depriving others who have the misfortune of being located downstream. Can management participation by the water users help solve this well-known situation? Or would the withdrawal of state authority actually encourage powerful farmers to take even more water, with no fear of government sanctions? Social theory suggests that under user management, tampering with water distribution becomes much more difficult. There is greater transparency and accountability rendering illicit water deliveries more difficult to conceal and even more difficult to defend. Whereas thefts of "government" water might be tolerated by other farmers, when the water supply is allocated to the collectivity of farmers, any theft of water implies stealing from a fellow farmer, and a fellow member of the association.

Lack of Incentives for Saving Water

What are the incentives that can encourage farmers to conserve water? Can such incentives become operative more readily under user management or under government management? The set of factors that determine water conservation behaviour are complex. One factor is security of supply. If farmers know when they expect to irrigate next time, they may be more modest in applying water, whereas if they are afraid they might have to wait a long time for the next irrigation, they will be apt to over-irrigated as insurance against this uncertainty. Another factor may be information and management confidence. If farmers are aware that there is an overall scarcity of supply and they believe the scarce water will be equitably distributed, they will may be more likely to limit their own consumption, or at least refrain from taking more than their share. A third factor is the transferability of water rights. If farmers in one part of the command area do not need water this week, but may need extra water after three weeks, can they forgo water now and use it three weeks from now?

While these three types of incentives — security of supply, information and management confidence, and transferability of water rights — can be found under both government management and user management, they are more likely to be found under user management. The reasons have to do with accountability. When the manager are accountable to the users, those managers have a direct incentive to keep supplies regular, and to inform the users of any problems. The continued employment of the managers depends upon the management confidence of the users. Similarly, the transfer of water rights from one user to another, or from one association to another, is not impossible under government management, but it is unusual and perhaps non-existent. Water markets are the formal manifestation of transferring water rights. While water markets are very much a novelty in the irrigation sector, there is a promising potential for tapping market forces to provide economic incentives for saving water, and for using it in the most productive manner.

Poor Drainage

Investments in drainage are notoriously under-valued in government managed irrigation systems. The emphasis is nearly always on increasing the supply of irrigation water, rather than on facilitating disposal of drainage water. It would be unrealistic to claim that farmers are necessarily more foresighted than government agencies in giving attention to drainage, but there is evidence from Mexico that more attention is given to drainage after transfer to user management. Part of the reason in the Mexican case is that the user-managers are able to operate the irrigation system at an overall lower cost, and therefore have funds available for drainage O&M.

Another part of the reason is that the government agency continues to subsidize drainage through monitoring soil and water quality and helping the new managers address priority problems strategically. In Mexico the water user organizations are responsible for both irrigation and drainage. In the Chambal project in Rajasthan, India, separate drainage committees are being established to operate and maintain new investments (by the government) in on-farm drains.

Rationale for Participation

Why participation? Another question might also be asked: "Why should the government be involved in irrigation?" Clearly, there are investments that only the government can make, or where the government has a definite advantage vis-a-vis farmers, even very well organized associations of farmers. Construction of dams and barrages, for example, or large canals, would be extremely difficult for farmers to handle. Governments provide us with available institutional resources — departments, agencies, trained staff, etc. — which can be used to get things done. Why re-invent the wheel and ask farmers to organize their own arrangements for building as dam?

- Comparative advantages
- Improved design, construction, and O&M.
- Lower costs to government.
- Social capital.

Comparative Advantage

Farmers have some comparative advantages as well. They have direct incentives to manage irrigation water in a productive and sustainable manner; they offer an on-the-ground presence that even the most dedicated off-site agency staff cannot equal, and they have an intimate knowledge about their fellow irrigators. The logic of the PIM approach is that both governments and farmers have separate comparative advantages. At the moment, governments are trying to do much more than they can do well. What are the advantages that management by farmers — by the users — can offer?

Improved Design, Construction and O&M

When farmers are directly involved in the design process, whether for new systems or rehabilitation of old ones, they will provide useful design input and they will come away with an understanding of the design logic of the system they will be managing.

During construction, farmer input has the functions of quality control (ensuring design standards are met), cost savings (through guarding against needless spending, and substituting some costs with farmers' own labour), and construction knowledge. Knowing how the system is constructed will help in repairs later on. The advantage of farmer inputs into O&M, either as direct managers or as the overseers of technical managers, has been discussed.

Lower Costs to Government

Cost savings to the government irrigation agency is often the driving force behind irrigation policy reforms. Government run systems are chronically short of maintenance funds leading to deteriorating systems and more difficult operation. Management transfer of major levels of the system to users offers government agencies an escape from this vicious cycle. While some critics see this as merely passing the costs on to farmers, the picture is not usually so bleak. Evidence from Mexico and Turkey suggest that farmers can manage better and more cheaply than their government predecessors. Thus, both farmers and government can benefit from these cost savings; farmers can enjoy better service, and cost savings; the government incurs less management cost and can then afford to improve service in the main system.

Social Capital

The organizations that farmers establish for managing their irrigation systems constitute a form of social capital that can have spin-off effects in other aspects of social and economic life. The network of contacts among agency staff and the water user organization leadership, for example, can bring the farming community into closer touch with related services, e.g., credit, educations opportunities, or even political access. And the skills that farmers learn through their experience with their water user organization — accounting, budgeting, planning, organizing — constitute a set of knowledge that can be used in many other productive endeavours.

What is PIM?

- PIM means "Participation" — not only in O&M and financing, but in making decisions that will affect the O&M — and financing
- Continuum of management approaches and levels of participation
- The concept of users' organization and transfer of management to users is different from PIM.

PIM Means Participation

Participation refers to a continuum of involvement in management decisions. One meaning of "PIM" may be that the irrigation users have total control and responsibility over the operations and maintenance of part or all of the irrigation system. Another meaning of PIM may be that a farmer council plays an advisory role, with real power remaining in the hands of the irrigation agency. Various levels of participation are outlined below.

1. Information sharing
    - Translation into local languages and dissemination of written material using various media
    - Informational presentations and public meetings.
2. Consultations
    - Meetings
    - Field visits and interviews.
3. Joint Assessments
    - Participatory assessments and evaluations
    - Beneficiary assessments.
4. Shared decisionmaking
    - Participatory planning
    - Workshops and seminars to determine positions, priorities, roles
    - Meetings to resolve conflicts, seek agreements, engender ownership
    - Public reviews of draft documents.
5. Collaboration
    - Formation of joint agency/stakeholder committees/task forces
    - Joint work with user groups, NGOs, or other stakeholder groups
    - Stakeholder groups given principal responsibility for implementation.
6. Empowerment
    - Capacity building of stakeholder organizations
    - Hand-over and self-management by stakeholders
    - Support for new, spontaneous initiatives by stakeholders.

Continuum of Involvement in Management Decisions

We can characterize the range of state-user relationships as a continuum from the state doing everything on behalf of the users, to the case of the state doing nothing for the users, other than leaving them alone. In between these two ends of the continuum is a very large gray (or blue) area where a government agency performs some management functions and farmers perform other functions. For purposes of discussion, we can divide the continuum into four types from more to less government involvement:

Type 1: Government does everything. In Malaysia, the Department of Irrigation and Drainage provides for the operation and maintenance of the main and secondary canals, while government sponsored farmers' organizations are responsible for providing water to individual farms. Farmers have no responsibility, and make no management decisions, about the water upstream from their outlets.

Type 2: State dominates; users help. The conventional management division in large irrigation systems is that the state takes responsibility for operation and maintenance of the head works such as a dam or river diversion, and the main, secondary, and larger tertiary canals, while farmers are responsible for managing water distribution and maintenance along the lowest level canals. Typically this entails farmer groups of between 10 and 50 farm families who are expected to work out sharing arrangements on their own.

Type 3: Users dominate; state facilitates. In some countries, associations of water users enter into contractual agreements with state water agencies for the provision of specific water services. In the case of Mexico, the National Water Commission manages the head works and main canals, while legally recognized water user associations employ their own technical staff for the management of the secondary and tertiary levels of the canal networks. Farmers pay their associations for the water, and a small portion of that fee is passed on to the National Water Commission for their services.

Type 4: Farmers do everything: In the Hill regions of Nepal most of the irrigated area is in the hands of local communities who have constructed their own canal systems, generally tapping small stream flows. Similar examples of local, farmer-managed systems can be found in nearly every country where irrigation is important, and the rules and customs of such systems provides a valuable pool of local knowledge that can be tapped in developing new irrigated areas.

Users' Organizations Different than PIM

Management approaches in irrigation generally fall into three categories: (1) public sector management, (2) private sector management, and (3) users' organizations. This last type can be termed "userism," and the process of transferring management to users can be termed "userization". The concept of "userism" is quite different from "privatization" in that we are talking about transferring management not to a third party "owner" who would purchase the irrigation system from the government and then hire out irrigation services to farmers. Rather, the PIM concept is more akin to an employee owned business that gives equal shares. Countries can be ranked according to their level of "userization" in the irrigation sector. While the ranking is rough, the trends are real.

At the upper end of the "userization" graph we find the United States, France, and Japan. Irrigation users have largely replaced the state in managing the irrigation sector although the government retains regulatory functions. At the lower end of the scale, where the state continues to dominate most aspects of irrigation and down to the tertiary or even quaternary levels, we find Morocco, India, Pakistan, etc.

The strategies that countries have taken in implementing PIM policies may be characterized according to three basic approaches: (1) the rapid "big-bang" approach of Mexico where water users are strongly pressured to establish an organization to replace the government, (2) the "bottom-up" slow approach of the Philippines with a strong focus on organizing and consensus building, and (3) a hybrid approach which adopts a moderate pace, such as that adopted by Turkey.

Is participation always necessary? Doesn't participation interfere with efficient management in some circumstances? Do we have to allow farmers to come into our board rooms and advise us on how to do our jobs? Aren't there some natural limits to what irrigation professionals should be responsible for and what farmers should become involved in?

A good rule of thumb is that a participatory dimension is important to all management functions. Perhaps there are exceptions to this general rule, but within the field of irrigation management, it is difficult to imagine any. This does not mean that a farmer's council has to be consulted before any decision is taken. If the water availability is so small that only 40% of the demand can be met along a given canal, do farmers need to be asked if they want the water? However, the farmers who receive only 40% of their demand do need to know about overall water availability so they can plan their response, and perhaps suggest better ways of utilizing their reduced share.

"Userism" as a Management Type

We may broadly classify management relationships into three kinds: the first is public management such as the irrigation department. The second is private management such as the Continental Corporation which produces Sparkletts mineral water. A third type of management is neither public nor private in the usual sense. We may call this a users' entity, such as a water user association. To describe this type of entity, we may use the term, "user-ism". You will not find this word in any dictionary; it was coined by Mr. Asif Kazi, Special Secretary in Pakistan's Ministry of Water and Power. We have adopted the term because it captures in one word the process of transferring management from the public sector to organizations of users.

Among these three basic types of management, as applied to irrigation systems, the most rare type is private management. This is mainly because irrigation water is a social good involving large numbers of small farmers, and it is very difficult for a commercial company to manage it with profit. This type of management is clearly not a general option for the irrigation sector. What about management by the public sector? While this is the most common type of management that we see today, in most cases public management has low efficiency and requires substantial subsidies. Experience from many sectors, including irrigation, tells us that it is almost impossible to bring public management into high levels of efficiency. Certainly it is possible to improve the management of public irrigation systems, but it is an uphill battle. The interests of the public managers are unlikely to coincide with the interests of the actual users.

The remaining management option is management by users, or participatory irrigation management (PIM). Under this situation, the managers have a direct incentive to manage the irrigation system efficiently because they are themselves users or are directly accountable to the users. This is the logic of userism: we can ensure a coincidence of interests between managers and users because the users are themselves the managers, or the employers of the managers.

Implementation Strategy

The opportunities for participation are different in each phase of the project cycle. Much of the emphasis on PIM has focused on participation in O&M, and particularly in the recovery of O&M service fees on behalf of the irrigation agency. While this aspect of participation is of great practical importance, there are many ways other aspects of irrigation management where participation can be incorporated. These include: (1) participation in irrigation project identification, planning, and design; (2) participation in system layout and construction; and (3) participation in project monitoring and evaluation.

In short, any aspect of irrigation management can have a participatory dimension. We have discussed why participation us important. In this section we will consider how to achieve it: how to implement participatory irrigation management. There is no recipie for this; indeed, the process of formulating a strategy that fits the specific features of any given country is the first — and ongoing — step. There are some common issues to consider, however, of which we will discuss two:

(1) creating an enabling environment, and

(2) start-up, pilots, and expansion phase.

Creating an Enabling Environment:

- Willingness and interest of stakeholders
- Strengthening interest and commitment
- The role of a PIM "promoter"
- Building consensus about the need for change
- Building Consensus among Farmers
- PIM as one part of a broader "package".

Interest and Willingness of Stakeholders: For participation to work, the government, the incumbent power broker and major "stakeholder" in most national irrigation sectors, must be willing and interested. At least three sections of the government must be willing and interested to support PIM: Political leadership, Administrative leadership, Irrigation agency leadership.

A second, and some would argue more powerful stakeholder must also be brought into any discussions of PIM policies at the earliest stage. We refer here to the farmers themselves, who have the most to lose — and also to gain — from changes in the way their irrigation systems are managed. The membership of farmer stakeholders overlaps with governmental stakeholders in the form of political leaders who represent farming constituencies. These political forces who can speak on behalf of both government and farmers, can be particularly important in both designing and promoting PIM reforms.

Strengthening Interest and Commitment: Expressions of interest in PIM on the part of these entities can be strengthened by facilitating participation in workshops where international experiences are shared (e.g., WBI seminars in Mexico, 1994; Turkey, 1996; Tokyo, 1997; IIMI/FAO seminars in China, 1994; Thailand, 1996). Specific study tours to "model" countries and schemes can be arranged. The World Bank had arranged such study tours to Spain, Mexico, and recently to Argentina and Chile.

Teams of political and administrative leaders have participated in these visits. The first such visit to Mexico by Turkish senior policy-makers was followed by study tours of several contingents of government officials from the Public Works Department. These visits and the shared experiences clearly contributed to the speedy implementation of the PIM program in Turkey. In Egypt, a tour of the USAID-assisted Irrigation Improvement Project was organized for legislators by the Ministry of Water Resources. The tour contributed to a much better understanding of, and subsequent policy support for, shared financing and management of the irrigation system by the government and water users. Methods and tactics for promoting PIM is explored in detail in another section of this Handbook.

The Role of a PIM "Promoter": One clear lesson from participatory experience is the importance of a PIM promoter or "champion" who is effective in mobilizing support within the government/irrigation agency. Typically this role is played by a key official within the agency. In the Philippines, for example, the Assistant Administrator of the NIA is rightfully considered the father of participation in that country's irrigation sector. In Mexico, there were well placed champions within the national water agency (CNA). Perhaps it is possible to rely on a senior consultant (who has the ear of top officials) to mobilize support within the agency. This is the approach being tried in Orissa, India, where a consultant on "farmer organization and turnover" will serve as a guide, but not a direct supervisor, to irrigation department field staff re-trained as organizers. A cadre of specialized social organizers will also be involved as assistants to these engineers. The job profile of the consultant on farmer organization and turnover is given in Annex One.

Building Consensus about the Need for Change: Stimulating a policy dialogue about the need for change and options to be considered can be based both on experience within the country, and by examples from outside the country. Within the country, public discussion of problems faced by farmers whose water services are unreliable or inequitable could be complemented by highlighting other cases where management reforms have resulted in improvements. Universities and centres of higher learning can help organize public discussion events. The media can help in reporting the outcome of these events and highlight key issues in the debate. NGOs can assist in seminars with water users. In Egypt, a professionally-made video film was made available to television stations that attracted large rural audiences in a pilot project area.

Bringing policy makers into contact with PIM cases in other countries is a high priced activity, but can be a powerful ingredient in swaying long-held opinions. Study tours, if carefully arranged and if the right people are involved, can make dramatic differences in the outlooks of individual officials. When Turkey was considering reforms

Building Consensus Among Farmers: A danger in designing and implementing PIM programs (and a cause for their failure) is the lack of attention to farmer interest and support. Since the impetus for PIM oftentimes originates in the perilous state of irrigation agency budgets, the focus for PIM from the agency's perspective could be quite limited, e.g., targeting financial contributions from farmers. The question, of course, is: why should farmers be interested in organizing themselves, if only to pay higher fees? Improvement in services and potential for income enhancement are better motivators for user organization. Farmers' incentives cannot be assumed, but rather must be assessed through field interviews and discussions with a representative sample of the community concerned. Farmers' involvement in designing a management model for PIM which builds-in strong incentives is a critical and often neglected step in the overall PIM implementation process. These issues are explored in another section of this handbook.

PIM as one Part of a Broader: Farmers are interested in much more than just the irrigation system; they want to improve their agricultural production, and more broadly, to improve their livelihoods. In Mexico, when the National Water Commission first met with farmers to discuss their interests, and what their priorities would be for irrigation improvements, the Commission expected that the farmers would ask for more lining of their canals. This is an expensive but important improvement to canal networks in many areas. The Commission was surprised to hear from the farmers that what they were most interested in was full management autonomy. Instead of a physical improvement, they were looking for an institutional improvement. Both types of improvements were within the authority of the government to arrange, but through different means; one required a simple financial outlay; the other required legal changes.

Start-up, Pilots and Expansion Phase

The design of an implementation strategy involves planning for the start-up, piloting and expansion phases of a PIM program:

- *Start-up.* To begin with, management responsibility for PIM must be assigned at the strategic management as well as the operations management levels. At the strategic level, a steering committee could be formed of key officials in the respective ministries, say, public works/irrigation/water resources, agriculture, planning, and finance. This group would approve of the PIM program — its goals, strategies, and a specific work program and budget on an annual basis. At the operations level, responsibility for PIM would be assigned to a specific senior manager. Experiences as to who should be assigned the responsibility vary among countries. In some cases, a senior manager is designated exclusively for Farmers' Organization. In others, a senior manager for Planning or O&M is assigned the additional responsibility. There is no formula for the assignment but whomever is chosen should be interested and committed to PIM ideas. This senior manager would need a group of interested staff from various levels, and representing social science as well as engineering, to help him implement the program.
- The start-up phase would also need resources for preparatory work for the PIM program. In-house expertise may be adequate for this purpose, but often external consultancy inputs may be required. A budget to prepare a pilot three-year program would be a good start and would have to be secured. Sources of financing could be the government's own funds or other external agencies such as bilateral and multilateral donors.
- Piloting. PIM is usually not a new concept to the region or the country. Farmer managed systems exist in most countries. Nevertheless, what is usually new is PIM in all stages of the project cycle, from planning and design to construction and O&M or PIM in the form of irrigation management transfer. In this context, it is useful to talk of "experimenting" with or piloting the PIM idea to test the appropriateness of the various PIM elements to local conditions in the country.

- Yet because PIM implies a new paradigm for the government's relationship with farmers, piloting is not possible in the conventional sense of the term. The irrigation agency must establish a commitment to the basic principles of PIM as a precondition for success in the pilot. The purpose of the pilot should be to help inform the agency how best to operationalize PIM policies, and to demonstrate what kinds of PIM approach would be feasible. It should not be seen as a "wait-and-see" test of whether PIM is a good idea. Inevitable, there will be an internal demonstration function of any PIM pilot, but the management of the agency, at the very least, should support the concepts of PIM at the time the pilot is launched.
- ? The objectives of piloting are to learn from experiences on a small scale so that a manageable irrigation system or subsystem can be the focus of implementation, monitoring, and learning. The implication is that the pilot projects should cover a range of conditions and be carefully monitored so that changes can be introduced in the original PIM model. For instance, in Egypt, 6 pilots covering about 70000 acres were launched in various parts of the country in the late 1980s. These experiences were evaluated in the early 1990s so that the lessons could be used in another set of irrigation schemes serving a command area of about 250000 acres.

Selection of pilots needs careful consideration. Selection criteria would normally include:

- interested farmers in the area
- relative lack of conflict among farmers or a keen desire to end the conflict
- water delivery system that functions relatively well
- tangible improvements can be demonstrated in a relatively short period of time
- supportive irrigation O&M field staff enjoying good rapport with farmers
- local government officials do not oppose the pilot
- size of the pilot scheme is neither too large (and therefore unmanageable) nor is it too small (results have no visible impact on the rest of the government-managed schemes).

To ensure their relevance, pilots must be part of a phased program that includes start-up, pilots, and expansion. Thus, the Mexican program aimed at the management transfer of all of its large scale irrigation systems (below the secondary canal level) covering nearly 3 million ha in a phased program, learning from experience during implementation. All of the phases should ideally be designed before pilots are implemented, thus placing them in the overall context of a country's institutional reform program. An additional problem with pilots is that they attract a great deal of resources, sometimes disproportionate to the size and scope of the experiment. Replicability becomes a casualty.

Expansion: At the end of the pilot phase, policy makers and managers would have derived lessons regarding the scope for user participation in the management of water, funds, operation and maintenance and capital development of the system. They would also be wiser about current government policies and procedures that help or hinder effective participation. The period before the large-scale expansion phases presents an opportunity to bring about changes in existing laws, policies, financing arrangements and procedures as the need may be. Without this intervention, an expanded program could run into severe problems. What was possible in an informal, small scale setting during the pilot phase may not be easy during the expanded large scale phase.

In addition to required changes in current policies, agency managers have to face the question of institutional capacity — trained staff, budget, managerial resources — to undertake a large scale expanded program that would cover a larger geographical area. Some time may have to be initially spent during the expansion phase in strengthening institutional capacity to manage the effort. These issues will be further explored in the section on Training Strategies.

Training Strategy

What Training do the Organizers Need?

Assuming that agency staff will be used for the organizing, and assuming that their professional training has been in irrigation engineering, they would need to be re-tooled as organizers. First of all they would need a thorough understanding of the rationale for PIM, so that they can present a clear message both to the farmers and to their own colleagues within the irrigation agency. Secondly, they will need training in communication skills (including listening skills) for effective interaction with the users. Thirdly, they will need training in social analysis, including an understanding of social stratification (by caste, ethnicity, or class), kinship, patron-client relations, labour relations, religious factors, political affiliation, land tenure (tenants, share-croppers, owners), etc. And fourthly, they would need training in methods for gathering information from farmers (e.g., participatory rural appraisal) and in methods for organizing farmers.

Legal Framework

The legal framework for the establishment of WUAs, and for enabling them to operate and maintain such parts of the irrigation system, consists basically of three sets of legal instruments, namely:

- The enabling law,
- The bylaws of the WUA, and
- The transfer agreement between the irrigation agency and the WUA.

The existence of a basic law on water, or on WUAs, is certainly an important parameter for the other parts of the legal framework because the basic law would usually specify the main issues that need to be included in the bylaws and transfer agreements, and would also determine the manner in which those issues are addressed. Those issues would usually include the procedure for establishing WUAs, the rights and duties of the WUA and the irrigation agency and the relationship between them, and the structure of the water rates and other fees.

The Enabling Law

For a WUA to be established as a legal entity, there has to be a law authorizing its establishment. This law could be a general comprehensive "Water Law" that deals with all aspects related to water, including establishment of WUAs. The National Water Act in Mexico, and the Water Resources Act in Nepal, are examples of such a comprehensive law. The enabling law could also be special rules and regulations dealing specifically with WUAs, and deriving their authority from a basic law, such as the "Implementing Rules and Regulations on the Provisions of Republic Act No. 7607" on small farmers in the Philippines.

Because of the absence of a basic law specifically on water or on WUAs, the states of India have relied on different laws to establish WUAs. In the Indian State of Maharashtra, WUAs have been established and registered as co-operative societies under the "Co-operative Societies Act." On the other hand, in the State of Tamil Nadu and the State of Orissa in India, WUAs are established and registered as societies, under the "Societies Registration Act."

The law establishing WUAs would usually include provisions indicating that the WUA to be established is a legal entity. Such enabling law would also address the relationship between the WUAs and the irrigation agency, the duties and obligations of the irrigation agency, and those of the WUAs, and the structure of water rates and the operation and maintenance and other fees. The enabling law may also lay down some of the main issues to be addressed in the bylaws of the WUA, and in the transfer agreement.

All societies, including WUAs, registered in any state in India under either the Co-operative Societies Act, or the Societies Registration Act, are legal entities capable of contracting, opening and operating bank accounts, and instituting and answering suits. However, because of their general nature, those two Acts do not address certain WUA related issues that are usually addressed in special water law or through a law establishing WUAs. It is for this reason that specific legislation dealing specifically with WUAs is often desirable.

Bylaws of the Water User Association

Whether established under a separate law or under an umbrella enabling law, the WUA would normally be required to prepare and agree on its bylaws before it can be registered as a legal entity. Those bylaws may be called "Regulations," "Constitution," "Charter" or "Articles of Associations." The issues that such bylaws need to address include:

Basic Facts About, and Objectives of, the WUA: The basic facts would include the name of the WUA, the law under which it is registered and its registration number, its address, and a clear definition of the area that the WUA is serving or its area of operation. This area of operation could be an entire irrigation district, or an entire command of a distributary, minor, sub-minor or a water course. It could also be defined by its size in acres or hectares. A broad statement on the objectives of the WUA is usually included in the bylaws. Such objectives would include: participation in the management, operation, maintenance and upgrading of the irrigation infrastructure works that the WUA has taken responsibility for, collection of water charges, and provision of irrigation and drainage services to the members of the WUA.

Criteria for Becoming a Member of the WUA: Most bylaws restrict membership of the WUA to the registered land owners in the hydraulic unit, who are engaged on a full-time basis in farming. If any member of the WUA sells his land, his membership will be automatically cancelled, and the new owner will be eligible for the membership of the WUA. However, the bylaws in some countries extend the right to become a member to both owners and tenants. In the Indian State of Maharashtra membership of the WUA is extended to "Any owner/cultivator/permanent tenant/protected tenant in the area of operation of the society...." These bylaws extend membership of the WUA beyond owners and tenants to include other categories of users such as sharecroppers and encroachers who are prevalent in some parts of India.

Number of Farmers Required for the Establishment of a WUA: Most of the bylaws state that at least 51% of the registered land owners in the command area where the WUA would be established should be enrolled as members before the WUA can seek registration, and before it can be allowed to operate. However, some bylaws allow the WUA to seek registration based either on the number of farmers enrolled, or the size of the land holdings coming under its operation regardless of the number of farmers enrolled.

In Orissa "At least 51% of the registered land owners in the command area covered by the Association should be enrolled as members, or the land holdings of the members should cover at least 51% of the total area under the proposed Association." Moreover, in some cases, once 51% of the farmers are enrolled and registered in the WUA, all other farmers within the command area will be deemed to have become members in that WUA. In Mexico "all users listed in the register who, while not founder members of the Association, apply and pay for irrigation services, thereby tacitly agreeing to belong to the Association, shall also be deemed to be members with the same rights and obligations."

The WUA as a Legal Entity: Although the enabling law would usually specify that the WUA is a legal entity, further details regarding what this entails are usually included in the bylaws. It is usually stated that the WUA is authorized to enter into contracts in its name, and that the WUA can sue in its own name, and answer suits instituted against it. The WUA can also be authorized to borrow funds from private sources, using, if necessary, its assets as a collateral.

Structural Organization and Internal Management: Although WUAs may be organized differently, the two most common ways of organizing a WUA at the membership level are:

(i) a general body of the WUA which would consist of all registered members who are current in the payment of their dues, as in

Mexico, Nepal, and the Indian states of Maharashtra, Orissa and Tamil Nadu; or

(ii) a general body, which could be called the general assembly, which would consist of delegates directly elected to represent the different irrigation districts or sub-units within the hydraulic unit, as in the case of some WUAs in Turkey. In the latter case, there is an absence of one single forum encompassing all members.

The executive body would usually consist of a president (or chairman), a vice president (or vice chairman), a secretary, a treasurer, and other specified number of members, and those posts do not, in many countries, carry remuneration. One exception is Turkey where the president of the WUA is paid a salary, and the other members of the executive body are paid honoraria.

Membership of the executive body may also be extended to non-governmental organizations (NGOs) or any other institutions interested in irrigated agriculture, including a representative of the irrigation agency, but such members usually do not have the right to vote.

A bank account in the name of the WUA is usually opened, with separate sub-accounts for water charges, operation and maintenance fund and membership fees. The executive body would also approve work expenses, engage labour, organize labour contributions from members, keep systematic accounts and records of amounts collected and those spent on the water charges, operations and maintenance fund, and membership fees. The executive body is usually required to have such accounts and records audited annually, and submit such accounts and audit reports to the general body during the annual meeting for approval. The executive body could include members who are specifically designated as auditors.

Operation and Maintenance: One of the primary objectives of the WUA is to operate and maintain the transferred irrigation and drainage system efficiently and economically, and with the full and active participation of all the members. Operation would include receiving water in bulk from the irrigation agency at a prescribed rate at the head of the minor/distributary and distributing such water equitably and in a timely manner, as per procedures and criteria agreed with the irrigation agency, to all farmers in the hydraulic unit, whether members or non-members. The bylaws would lay down, in agreement with the irrigation agency, the criteria for allocation of water to both members and non-members, which could be based on the type of crop grown or the size of the area to be irrigated, or both. The bylaws would also include the criteria for assessing water charges and operation and maintenance charges from both members and non-members. The operation and maintenance fund could include sources other than charges from farmers.

Water Charges: Reference to water charges is usually included in the enabling law. Other details on water charges are also included in both the bylaws and the transfer agreements. Inclusion of provisions on water charges in the bylaws would serve the purpose of establishing the payment obligations of each member of the WUA, whereas the provisions in the transfer agreement would establish the payment obligations of the WUA, as a legal entity, vis-a-vis the irrigation agency.

The transfer agreement would usually include provisions on the manner in which water charges are calculated, and the due date or dates for payment of the water charges by the WUA to the irrigation agency. It may include provisions for the payment of commission or discount by the irrigation agency to the WUA on water charges collected by the WUA. The transfer agreement would also include provisions giving the irrigation agency the right to suspend delivery of water if the WUA fails to make the payments of the water charges within the prescribed or extended time limit. Moreover, non-members of the WUA may be required to pay higher water rates than those paid by members.

Rights and Obligations of Members: Under most WUAs, each member of the WUA has one vote regardless of the size of his land holding. The bylaws would also need to address the issue of proxy voting-whether it is allowed, and if so, the maximum number of proxy votes one member may cast on behalf of other members. Failure of a member of the WUA to meet his membership obligations as described in the bylaws, such as failure to make payments, permit inspection of the irrigation system in his land, comply with the terms of the transfer agreement, carry out proper maintenance, or allow delivery of water to other users may subject such a member to sanctions. Such sanctions may include suspension of such a member until all outstanding obligations are met.

Interpreting and Amending the Bylaws: Provisions would usually be included in the bylaws themselves describing the procedures for interpreting provisions of the bylaws in case there are different views as to what a certain provision may mean. Procedures and quorum required for amending the bylaws would also be included in the bylaws. Usually interpretation of provisions of the bylaws would be referred to a central body such as the Registrar of Societies, or the irrigation agency, and amendments would be effective after approval of such body.

Establishment of a Federation of WUAs: The bylaws of some of the WUAs, as in Mexico, Colombia, Tamil Nadu and Orissa, include provisions enabling the establishment of a federation encompassing registered WUAs in one command area. Those bylaws specify the responsibilities of the federation, and the relationship between such a federation and each of the affiliated or member WUAs. Usually, the presidents of each of the WUAs that decided to establish or join such a federation would represent their WUA in that federation. The federation would have an advisory non-binding role over the member WUAs, and may be used to resolve any disputes among such member WUAs, or between WUAs and the irrigation agency.

Bylaws of the Water User Association

The transfer agreement is the agreement between the WUA and the irrigation agency in which the irrigation agency agrees to transfer to the WUA responsibilities for operation and maintenance of certain parts of the irrigation system, including the drainage system, and the collection and remitting of water charges; and the WUA agrees to carry out such responsibilities. This agreement may also be called "Memorandum of Understanding (MOU)," "Transfer Protocol," "Concession Agreement" or just "Concession." The issues that such transfer agreement would need to address include:

Area and Irrigation System to be Transferred: The agreement would need to define clearly the irrigated area to be transferred, specifying the size of the area, and the command under which it falls, and including the irrigation system existing there that is being transferred. The system to be transferred is usually the irrigation system at the level of primary, secondary and tertiary, including the drainage of such areas too. A copy of a map showing such area may be attached to the agreement. The agreement would specify whether the ownership of the irrigation system, including the land and structures and works there on, remains with the irrigation agency or is being transferred to the WUA, together with the operation and maintenance of such irrigation system. Provisions should also be included clarifying whether the ownership of any ancillary equipment is being transferred.

Interim Joint Management: Some agreements may provide for a joint management of the irrigation system for a short period of time by both the irrigation agency and the WUA. The rationale for such joint management is to prepare the WUA, during this interim period, for taking over full responsibility for operation and maintenance of such irrigation system. During this interim period which may run for up to one year, officers of the irrigation agency would train WUA representatives in compiling necessary data, and preparing and testing operation and maintenance plans for the distributaries, minors and subminors to be transferred to them. They would hold joint inspections to identify any problems in the irrigation and drainage system, and to agree on how to deal with them.

Transfer of the Irrigation System: Transfer of the irrigation system to the WUA will normally be preceded by a number of actions, including the preparation of an inventory of the works, structures and equipment to be transferred, joint inspections of those works, structures and equipment by both the irrigation agency and the WUA representatives, carrying out of necessary testing, and repairs, if any, at the irrigation agency cost, and handing over management of the system, along with all necessary documents and instructions, to the WUA. The WUA would need to satisfy itself that the system is, indeed, in a good working condition, as the transfer agreement would include provisions that the irrigation system, at the time it is transferred, was in good working condition.

Responsibilities of the Irrigation Agency: The agreement would spell out clearly the responsibilities of both the irrigation agency and the WUA. Responsibilities of the irrigation agency would include handing over the system in a reasonably operating manner, delivery of water to the WUA in bulk at the agreed time and providing the WUA with any agreed upon financial assistance and other benefits. The agreement could include provisions absolving the irrigation agency from liability should it be unable to deliver the agreed upon amount of water, or unable to deliver it at the agreed time, for reasons of force major, or act of God. In periods of water scarcity or emergency, after the demand for domestic and other priority uses is satisfied, the irrigation agency would usually have the authority to decide that the remaining water for irrigation shall be allocated to crops of utmost importance to the community there.

Responsibilities of the WUA: A number of the responsibilities of the WUA detailed in this section of the agreement are usually spelled out in the bylaws of the WUA, but may still be included in the transfer agreement to clarify the obligations of the WUA towards the irrigation agency. Such responsibilities would include: operating and maintaining the irrigation system transferred to it, including the drainage system, in a proper and satisfactory manner; receiving water in volumetric basis, and distributing such water equitably and in a timely manner, based on clearly defined criteria, to both members and non-member farmers in the operation area, and collecting the water charges agreed with the irrigation agency.

The responsibilities also include establishing the operation and maintenance fund, and maintaining and repairing, in a satisfactory manner, all the field channels, field drains, minors, subminors and distributaries, together with the structures there on in the operation area of the WUA. In addition, the WUA may be responsible for the maintenance and repairs of any equipment and machinery transferred to it. Such equipment and machinery may be transferred to the WUA as part of the irrigation system for which it is now responsible, or may be separately leased by the irrigation agency to the WUA at an extra cost.

The WUA would also be responsible for the security of the infrastructure transferred to it, and such responsibility could either be carried out by the members of the WUA themselves, or through hired labour. Maintenance would usually include silt clearance and removal of weeds from all water courses under the WUA. It would also include earthwork to restore banks and repairs to other structures, in addition to maintenance of service roads. Usually minor repairs are carried out by the WUA and major repairs by the irrigation agency. Definitions of what is "minor" and what is "major" should be included in the transfer agreement (repairs to damage caused by natural disasters such as heavy rains, floods or earthquake are usually considered major repairs). The WUA may be required to prepare an annual maintenance program for the irrigation system under its responsibility, including any machinery and equipment, and to submit such program to the irrigation agency for approval prior to implementation. Operation and maintenance of the irrigation and drainage system, other than the one transferred to the WUA, would continue to be the responsibility of the irrigation agency.

The agreement would authorize the irrigation agency to suspend supply of water to the WUA if maintenance and repairs were not being carried out properly, or to carry out the repairs itself and recover the cost from the WUA. The agreement would also include provisions on how disputes between the WUA and the irrigation agency, arising in the course of the operation and maintenance of the transferred irrigation system, would be settled. Such disputes could be referred to a committee comprising one representative from the irrigation agency and the water users' federation.

Termination of the Transfer Agreement: Although the enabling law may include provisions on the termination of the transfer agreement, usually more detailed provisions are included in the transfer agreement itself. The agreement terminates after expiry of the number of years specified in the transfer agreement, which may be as high as twenty years as in Mexico. However, the agreement would usually be subject to renewal for another similar period. Moreover, failure by the WUA to comply with the provisions of the agreement, including the failure to properly operate and maintain the irrigation system transferred to it, or to make timely payment of water charges, or to take corrective measures within a specified period of time, as agreed with the irrigation agency, would give the irrigation agency the right to terminate the transfer agreement.

Organizing Processes

PIM implies the establishment of an organization of water users. In cases of management transfer, the establishment process focuses on a level of user organization where no organization presently exists, that is, at a level of the irrigation network previously managed by the government irrigation agency. Typically this entails organizing federations of user groups at the tertiary level under a secondary-canal association. The challenge of creating a new organization of users is perhaps the most central feature of the management transfer process. The act of management transfer from the agency to the users depends upon a user organization that is capable of assuming those management responsibilities. Before any organizing of the users is carried out, there needs to be a package of incentives in place for both the users and the agency staff whose jobs would be affected by the transfer program. Such incentives are needed both to make the program work, and to maintain credibility with farmers whose long-term support will be required. If organizing is attempted before an adequate incentive structure is established, the transfer program could well collapse, thus setting the entire program back by several years, as well as causing short-term hardships to those concerned. Thus, if the incentives are not clear and attractive to farmers, the organizing process should be delayed until the incentives are clarified.

First determine:

I. Who will do the organizing?

II. What training do the organizers need?

III. Who will provide the training?

Then, the organizing steps for transfer of management to users.

Who will do the Organizing?

The first step is to decide on the type of organizers who will work directly with farmers in helping establish the organization. [We use the term "water user association" (WUA) to refer to this new organization, although the term, "WUA" can also refer to organizations at the tertiary level of the system which have always been outside the management control of the government irrigation agency.] In the Philippines, a special cadre of social organizers was recruited and trained by the National Irrigation Administration in the late 1970s and early 80s. These organizers were mostly social workers or social scientists selected for their ability to work easily with farmers in village conditions. They were trained in irrigation management so they could better understand the technical problems of the farmers they were trying to organize.

The work of the Philippines social organizers was effective, yet this approach has not been widely replicated by irrigation agencies in other countries. The investment and bureaucratic difficulties involved in recruiting new temporary staff from a different discipline has led many irrigation agencies to try other approaches. In Mexico's transfer program, the National Water Commission's own staff were used along with staff from a sister agency, the Institute for Water Technology (IMTA). In addition, a few consultants were brought in on a case by case basis. In India, some state irrigation departments have relied on extensionists from the Agriculture Department under the Command Area Development Program. Also in India, several NGOs have been involved in organizing, at the request of the irrigation departments. But the organizers of choice, for most irrigation agencies, will be their own field staff. These staff are already within the bureaucratic structure of the agency, so the lines of authority are clear, there is little additional expense involved, and these staff are already familiar with the physical systems and with the local farmers.

The problems with using irrigation agency field staff for organizing work, however, are considerable. These staff are not necessarily interested in organizing farmers, They have not received any prior training in social work (in most cases), and their superiors have also no training (nor interest) in these tasks. These would-be organizers must be re-trained for their organizing tasks, and just as importantly, their job assignments need to be redefined to reflect their new role. In addition, their superiors need to be trained and re-oriented so they understand and appreciate the new role to be played by their field staff.

What Training do the Organizers Need?

Assuming that agency staff will be used for the organizing, and assuming that their professional training has been in irrigation engineering, they would need to be re-tooled as organizers.

First of all they would need a thorough understanding of the rationale for PIM, so that they can present a clear message both to the farmers and to their own colleagues within the irrigation agency.

Secondly, they will need training in communication skills (including listening skills) for effective interaction with the users.

Thirdly, they will need training in social analysis, including an understanding of social stratification (by caste, ethnicity, or class), kinship, patron-client relations, labour relations, religious factors, political affiliation, land tenure (tenants, share-croppers, owners), etc.

And fourthly, they would need training in methods for gathering information from farmers (e.g., participatory rural appraisal) and in methods for organizing farmers.

Organizing Steps

Note: This assumes there are reasonable incentives for farmers to take over system, and for the government irrigation agency to hand over the system; if not, go back and work on the incentives!

- Organize the organizers: Arrange for supervision/support for the organizers, and clear lines of communication with Departmental staff responsible for the overall project (Ensure that PIM component of program is well integrated with rest of project).
- Meet the farmers and other irrigation stakeholders/Discuss plans formulated during participatory design phase:
  - — village head
  - — local administrative officials
  - — local political leaders (MLAs/MPs)
  - — leaders of other farmer organizations (producer cooperatives).
- Identify key power relations among farmers; develop strategy for organizing
- Establish provisional boundaries of the system (through consultation with key power brokers among the farmers); conduct inventory of potential members; draw map showing command area and irrigation system.

- Arrange series of meetings between farmers and Departmental field staff to discuss improvements that need to be made prior to handover.
  - arrange canal walk-through to discuss specifics of design/ infrastructure improvements
  - discuss general terms of WUA contracts.
- Arrange farmer visits to other associations to discuss with those farmers (and invite those farmers to visit new association).
- Organizational Assistance:
  - Help prospective WUA leaders arrange farmer meetings to discuss plans
  - Help formulate/revise bylaws
  - Advise on elections/selections
  - Assist with legal registration
  - Arrange management training for WUA leaders.
- Arrange meetings between WUA leaders and Department staff to discuss details of WUA contract and terms of transition phase leading to hand-over;
- Advise on staff recruitment(?)
- Assist with formal hand-over
- Visit periodically to monitor WUA's performance.

Reorienting Agencies

Agencies implementing PIM need to reorient themselves to a new style of water and infrastructure management. No longer is the agency the sole manager of the water system; rather, the agency staff become management partners with farmers. The agency retains control over the water supply and head works (in most cases) and perhaps the main canal network. Their management must be closely coordinated with the management of the lower levels of the system, now under the control of user associations. Under this new arrangement, the agency has both a direct management role (for the highest levels of the system) and an indirect management role — facilitating and supporting the work of the user associations.

- Restructuring the Irrigation Agency
- How agencies do NOT Change under PIM
- Facilitating and Hindering Factors.

Restructuring the Irrigation Agency

Even without an explicit PIM program, changes over the past two decades in the irrigation and water policy context clearly point to needed changes in agency roles and functions. Irrigation agencies are being forced — by citizen concern as well as by the conditions imposed by international financial organizations — to pay a great deal of attention to social and environmental impacts of irrigation activities. These have become important aspects of project assessments in the 1980s. Similarly, as user participation has gained importance, irrigation agencies have had to re-evaluate their roles and functions with respect to irrigation development and management.

Incorporating PIM requires a range of new functions and new organizational structures for the irrigation agency; it must shrink and shift to make room for user management. It must shrink in staff and functions, and it must shift from directly doing O&M to indirectly supporting O&M within the user managed level of the system. Specifically, changes in agency roles require the agency to perform new tasks as explained below.

- Revised Structure. At least some field-level staff of the agency, and perhaps most field staff, would be replaced by the new staff of the WUAs. As a general rule, higher level staff of the agency would be less affected by the transfer of O&M responsibilities.
- Consultation. Staff in the various functional departments of the agency need to commit to and gain expertise in farmer consultation.
- for those staff in planning and design, PIM implies a consultative step that would provide farmer inputs into prioritising system improvements, selecting designs and layouts, and determining outlet locations. On the part of agency staff and their consultants, consultation with farmers for planning and design requires listening skills and an ability to present technical details in simple language. It also implies the flexibility to change plans and designs based on farmers' suggestions.
- for O&M staff, PIM implies the ability to respond to feedback on system performance from water users. Field staff would need to reorient themselves from carrying on as operating engineers to performing advisory functions — a role in which they may have little experience or expertise.

- Accountability. The accountability of irrigation staff increases considerably with PIM and the formation of WUAs. Both the agency and the WUA need to maintain accurate records of water delivery (planned and actual) to ensure that the agreed amount of water is delivered/received. Where PIM is introduced at the survey and design stage itself, information on initial designs and comparative costs would have to be shared with water users. The importance of shared information increases significantly with PIM, since now agency staff are not only providing information "upwards" to their bosses but also "downward' to their clients, the managers of the WUAs.
- Advisory Services. Agency staff need to reorient themselves to extend technical and managerial advice to water users. PIM is effective only when informed water users participate knowledgeably in the planning, design, construction and management of irrigation systems. In Egypt, where water users are organizing themselves around a common pump at the head of the tertiary canal, the Ministry of Public Works and Water Resources has established an Irrigation Advisory Service to extend assistance to WUAs.

How Agencies do not Change Under PIM

Many of the irrigation agency's core functions continue after PIM is introduced. In most cases of PIM, the public irrigation agency maintains the legal responsibility for overseeing the wise use of the irrigation facilities. Under most contracts, the WUA is managing these facilities for a set period of time (25 year concessions in Mexico). Cases of outright privatisation — where the government agency permanently sells or cedes the ownership of the infrastructure to a WUA — are very much the exception. Thus, the public irrigation agency has not only a right but an active responsibility to regulate, for example, groundwater levels and soil and water quality.

Management functions also continue, with the agency maintaining responsibility for the head works and (usually) main system. The agency also has a monitoring role in ensuring that the WUA is adhering to the contractual agreement of management transfer, and is properly maintaining the infrastructure and adequately delivering water to the WUA members.

Facilitating and Hindering Factors

Incentives within the agency are often opposed to the establishment of strong WUAs. Organized users pose a threat to the agency's control and to the endemic problem (in many countries) of bribes and kick-backs on contracts. Management transfer programs which aim to expand the role of WUAs threaten the jobs of agency staff. And the task of mobilizing the formation of WUAs becomes an added responsibility for these same staff, and not a task in which they have either professional training or interest. Finally, because irrigation agencies are engineering organizations, the professional rewards and recognition are in design or construction of physical schemes, not in dealing with farmers' demands.

Given this situation, which naturally varies considerably from country to country, how could PIM possibly gain a foothold? How can an agency transform its own management culture? There are a number of tools for doing this:

- Performance evaluation criteria: Including work with WUAs in the job descriptions and performance evaluations of agency staff is one way to create incentives for PIM. Supervisors (whose own job descriptions also need to incorporate PIM criteria) need to monitor their subordinates' performance in working with WUAs, and request feedback from farmers on how helpful staff members are. Publicizing the performance plans of agencies among users may strengthen agency and staff accountability. While the number of associations registered is relatively easy to measure, it is not an adequate indicator of staff performance with WUAs because it does not take into consideration how well the WUAs and joint management work. Fee collection rates from farmers or from WUAs may provide a better indicator of farmer satisfaction.
- Monetary and non-monetary rewards: Monetary rewards are often difficult in the context of civil service regulations. In some cases, where the agency is an autonomous corporation, a reward system may be possible. The National Irrigation Administration, a semi-autonomous agency in the Philippines, was able to offer higher than regular civil service salaries to its staff. However, this was contingent upon maintaining a balance between expenses and cost recovery, which gave staff an incentive to work with WUAs to reduce agency costs and raise irrigation service fees. Non-monetary rewards can also provide a powerful motivator for agency staff. In Nepal, a special medal was awarded to an Irrigation Department engineer who had made extraordinary efforts to establish a WUA. Some already motivated staff will derive job satisfaction

from challenging new tasks such as WUA organization, provision of technical assistance, monitoring and regulation, rather than routine O&M. But the agency's management must provide clear and consistent signals that such new roles are valued.

- Reduced transaction costs. Reduction in the number of conflict cases (or "hassle factors") to be resolved by the agency staff has also proved to be a valuable incentive for working with WUAs. The establishment of WUAs as an organized forum for communication can reduce the transactions costs for agencies and for farmers. Reduction in damage to structures, as farmers focus on protection of system facilities eases the burden placed on agency field staff. Thus staff may have to devote less time to field inspections and user management often results in greater vigilance over infrastructure and fewer complaints. Thus, there may be less need to deal with individual farmer demands as WUAs take on additional roles. In Chile, problems with the state management of irrigation are no longer "politicized" by farmers since users' associations and organized forums for articulation of problems have been established.
- Skills Training. Preparation is needed for two types of activities that could result from the introduction of PIM:
- Activities dealing with user participation, including farmer organization, WUA establishment, technical assistance for WUAs, fostering agreements with informal and formal WUAs, public communication and information etc.
- Activities that can be taken up by agency staff as a result of water users assuming a larger role in irrigation O&M thus freeing up time and resources for the agency.

Agency management will have to formulate a program of short and long term training to prepare the staff for changes in roles and functions and upgradation of expertise. It may not be possible to meet all needs through training of available staff. New staff may have to be inducted into the agency. Secondment from other departments of government, from universities or NGOs may also be considered. For instance, in Egypt, the Ministry of Public Works obtains the services of agricultural extension staff through secondment from the Ministry of Agriculture. In Philippines and in India, community organizers are hired as consultants and contract staff.

Financial Aspects

O&M can be financed directly (through a direct tax based on O&M costs) or indirectly through land taxes or other kinds of agricultural taxes. The relative share of government vs. farmer contribution varies dramatically across countries, with some countries using O&M as a vehicle for delivering subsidies to irrigated farmers.

- How to set O&M cost
- User participation in defining service levels
- Assessing users' willingness and capacity to pay
- Determine appropriate charging mechanisms
- Linking fees and services.

How to Set O&M Cost

The managing agency needs to identify the various services it provides through the water delivery and drainage systems under its management control and then allocate costs to each service category. This information should be made freely available to farmers, and to their WUA leaders. Such information helps WUAs understand the respective contribution of government and farmers and helps the WUA determine which functions it might take over from government, and which functions it might prefer to pay government to provide. An aspect of the cost determination activity is the separation of "staff or administrative" costs and "works or program" costs. A clear rationale for the two types of costs is important in justifying the irrigation service fees which farmers are expected to pay. This is equally true for the water fee payable to the government agency, and to the fee payable to the WUA.

User Participation in Defining Service Levels

Involvement of farmers/WUAs early in project design to determine the level of service required by farmers is crucial. Users' involvement in setting service levels will ensure the genuine demand for those services, which will be expressed by a willingness to pay the fee required to support that level of service. Some examples of users working with the agency in determining what services the government should provide are given in the following:

- In Mexico, following the transfer of management of irrigation districts to WUAs, irrigation plans are prepared by WUA-hired managers at the level of the module, each covering about 5-8000 ha, taking into account cropping plans, conveyance losses, and equitable distribution. These are then negotiated with the national irrigation agency to determine the final allocation, generally based on an arranged demand pattern. O&M costs at the level of the secondary and below are met by user fees managed by the WUOs. In addition, WUOs contribute to a part of O&M costs of higher levels of the system.

- In the Office du Niger, Mali, joint committees have been established for O&M in every region. The committees have 5 to 10 representatives of producers and 5 to 10 representatives of the Office. These committees decide on types of services, costs including procurement matters, and water service fees. They also make decisions on the use of 50 per cent of the user fees collected for O&M.
- In Chile, the national federation of WUOs was consulted in the design of a Bank supported irrigation project. The federation and local WUOs played an active role in project preparation especially in discussions of service options and costs. Subsequently, the project incorporated the condition of WUO approval for investment proposals and other project components.

Assessing Users' Willingness and Capacity to Pay

The readiness of the user to pay will depend not only on the benefits to be derived, but on the level of payment to which the users have become accustom. It is in the farmers' interest to pay as little as possible. But it is also in the farmers interest to receive reliable supplies of water, and for this he will certainly be willing to pay something. In a rehabilitation project in Egypt which is introducing PIM, it has been estimated that users will derive incremental income from the project due to expansion of irrigated area and yield increases and that incremental costs of their participation in irrigation management would amount to about 20 per cent of this new income. Increases in income would also depend on agricultural extension services and these would have to be strengthened to allow farmers derive full benefits from the improved water services delivered by the project.

Financial viability of the WUA enterprise is critical to the sustainability of the organizations and the irrigation infrastructure. It is important to examine the total costs the WUAs have to bear, including staff, materials, travel to meet with government officials, and formal and informal payments that must be paid to government agencies. If the level of fees members must pay to meet these costs (which are often in addition to continued payments to the government) is too high a portion of gross or net income from irrigation, the WUAs are not likely to succeed. This is particularly problematic for pump irrigation systems, in which high energy and maintenance costs exceed what the organizations are able to collect from farmers.

In some cases, WUAs must depend on income from other sources to subsidize their irrigation activities. For instance, some irrigation districts in the Western United States rely on revenue from power sales to balance their irrigation budgets.

Determine Appropriate Charging Mechanisms

There are a number of alternative charging mechanisms available to agencies and WUOs in structuring O&M fees. As a principle, the fee structure should be equitable, administratively simple, and easily understood by both users of a particular service and the staff that will administer and collect the fees.

User Fees: Generally, user fees are charged by area irrigated. The advantage of this method is that it is relatively simple to understand. However, it does not reflect the quantity of water used nor does it provide an incentive to conserve water. Volumetric charges address those concerns. However, there are many measurement difficulties in operationalizing this method in the field. Agencies often favour a combination of the two methods. In Mexico, for instance, the agency charges the WUAs volumetrically at the turnout of the secondary canal. In turn, the WUAs base their water charges to individual members on area irrigated and type of crop. In addition to unit area and volume, WUAs have also resorted to other bases such as charges for the entire season which clearly favour the high volume user. The design of irrigation fees should be a topic in WUA training programs so farmers understand the implications of various modes of charging user fees for efficiency and equity.

Property Tax: This is a charge based on the value of the land receiving the service. Irrigation and drainage enhance land value. The higher the land valuation, greater the service payments. The problem with this method is that generally land taxes are collected by the Treasury and are not transparently linked with irrigation services. Also, the valuation and revaluation process demands a great deal of administrative resources.

In-kind Contributions: Contributions of labour, materials or both by the benefiting farmers towards O&M reduce the cost of services. In Viet Nam, the provincial and national governments finance schemes down to 150 ha for new irrigation development. Below this level, farmers must build the channels, with the government providing support for survey and design, as well as materials in some instances. After schemes are completed and taken over by farmers, each member must provide up to 20 person days per year towards maintenance of the tertiary as well as secondary systems.

Replacement of Assets: A significant cost in irrigation services is depreciation of the asset base used to store, transfer and deliver water services to users. Facilities employed range from minor tools and equipment, buildings and housing, motor vehicles and heavy equipment, canals and drains, pumping stations, and where applicable, major structures including dams. Unfortunately, depreciation costs are usually neglected in estimates of cost of water services whether provided by irrigation agencies or by WUAs. Assets are consumed at different rates and each year, a significant non-cash value is "lost" as assets depreciate, even with adequate maintenance. Without proper maintenance, service life will be shortened and depreciation will accelerate. The agreement between the irrigation agency and the WUA includes O&M standards to be maintained. Provision is also made for collecting and retaining disaster funds by WUAs. In Mexico, the Hydraulics Committee at the district level approves the annual O&M program proposed by the WUA. Care of assets is vital to the sustainability of water system performance. Maintenance and the ultimate replacement programs for all assets need to be planned and implemented in an effective and economical manner. In parts of Vietnam, the estimate of costs of services (and hence of user fees) include provision for depreciation of assets. Adequate data about the location, age, condition and serviceability of all assets is therefore essential to this process. WUAs need to be trained and assisted in preparation of an asset inventory and in the formulation of a maintenance program.

Linking Fees and Services

An obstacle to user contributions is the perception that their payments are not linked to the irrigation agency or its services. Often, money for services is collected by the central revenue department, and from there funds are allocated in the various government agencies, including irrigation, with little link between revenue earned and costs of service provision. To reverse this process, the money cycle must be transparent.

Ideally, WUAs would collect O&M fees and manage the system under their control and pay a service fee to the agency for that portion of the system under the agency's control. Under this arrangement, the agency and the WUA agree on mutual rights and responsibilities and the WUA is completely responsible for management of water, infrastructure, and finances for a part of the irrigation infrastructure.

If users are to pay service fees with the government managing the system, fee collection is preferably done through the WUAs. The Philippines experience has shown that fee collection is far better in systems where WUAs are organized and where they have a role and incentive for collection. For instance, in the pilot projects in Maharashtra, India, WUOs are allowed to retain a proportion of collections as a bonus. The cost of collection by WUAs is lower relative to government agents. Sanctions for non-payment by individuals are easier when enforced by WUAs. The danger is that considerable time and energy are devoted to fee collection on the part of the agency staff as well of the WUAs and fees rather than reliable water services becomes the focus of management attention.

Where users pay service fees to the managing agency, fees are a good means of signalling satisfaction with services, as in the case of Indonesia. Such information is publicized and used for evaluation of agency services and management. In any case, transparent accounts and audits are fundamental in showing WUA members that the financial management of the organization is sound. General body meetings of the WUAs must discuss financial performance including the accounts and audits.

WBI Training Programme

In most developing countries, irrigation development projects and their operation and management are heavily dominated by the public sector. Conventional wisdom has assumed that only the state was capable of handling large modern projects requiring heavy capital investment, complicated technical inputs, the legal mandate to distribute water, and collect fees.

Recent experience challenges these assumptions. Government-operated irrigation systems are often poorly maintained with steadily deteriorating infrastructure. Yet some of these same systems show dramatic improvement when their management is transferred to water user associations (WUAs), which enter into contracts with the government for operating and maintaining portions of the system or in some cases entire systems. Since the mid-1980s, countries including Mexico, Turkey, Indonesia, Philippines, Colombia, India, Srilanka, and Nepal have adopted policies to encourage greater management participation by water users. One of the most dramatic management transfer programs has been in Mexico, where the government adopted a policy to gradually transfer all its large scale irrigation districts to 78 WUAs. As of mid-1993, the management of more than 1.2 million hectares (ha) of irrigated land has been transferred to WUAs.

In response to interest in participatory irrigation management expressed both within the World Bank and in a number of borrowing countries, the World Bank Institute of the World Bank has initiated a five-phase program on participatory irrigation management (PIM). The overall purpose of WBI's program on PIM is to stimulate high-level policy dialogue on participatory irrigation management within the country, leading to policy commitment and programmatic action.

The PIM training program has thus far been initiated in India, Pakistan, and Morocco. Several other countries, including Egypt, Indonesia, Nepal, and several African countries are anticipated to join the program during 1996. Selection of countries is based on the relevance of participatory irrigation management to ongoing or planned World Bank loans and on the level of interest expressed by the host country and the concerned operational division.

For each country where it is offered, the program entails a series of three seminars (Phases 1-3) with the potential for further involvement in implementation (Phase 4) and evaluation/dissemination (Phase 5). The five phases are as follows::

Phase 1: A national seminar to introduce policy makers to the implications of PIM, consolidate national experience, learn about best practices from other countries, and formulate an indicative action plan for enhancing participation in the irrigation sector;

Phase 2: An international seminar held in a model country for PIM where participants can visit field sites to learn directly about the host country's experience and compare experiences with other participants;

Phase 3: A follow-up national seminar to formulate a national action plan for PIM;

Phase 4: Special-purpose seminars and/or training assistance to help support the implementation of a national PIM program; and

Phase 5: Support for evaluating national PIM programs and disseminating lessons learned to other countries.

Style and Methods

The seminars in phases 1-3 are designed to encourage the active participation and open discussion of all participants. Each seminar has the very practical goal of drafting an indicative or revised action plan for implementing participatory irrigation management. Presentations from international and national cases, and in particular the field visits, provide ideas to consider for including in the action plan. Much of the real work of the seminar is done in concurrent small group sessions where participants focus on a specific topic and discuss ideas in terms of their relevance to national conditions. The ideas that emerge from the small groups are presented to the plenary session for further discussion and possible incorporation into the indicative action plan.

Sponsorship and Cost-Sharing

Program activities are co-sponsored by WBI and one or more host country organizations, agencies, or ministries. The national co-sponsor normally covers all expenses for national participants (including air travel in the Part-II international seminar), with the understanding that funds can be drawn from ongoing World Bank loans. WBI generally covers the costs of international resource persons.

Participatory Irrigation Management: Benefits and Second Generation Problems

Acronyms and Definitions

Participation in irrigation management by water users can take a wide variety of forms. Farmers can be involved in various system management functions, including, planning, design, operations, maintenance, rehabilitation, resource mobilization, and conflict resolution. Moreover they can be involved in these functions at various system levels; from the field channel to the entire system.

Almost all irrigation systems have some involvement by water users in system management. When people speak of introducing "Participatory Irrigation Management" (PIM), they are thus usually referring to a change in the level, mode, or intensity of such participation that would increase farmer responsibility and authority in management processes.

Irrigation Management Transfer (IMT) is a more specialized term which refers to a process of shifting a number of basic irrigation management functions from a public agency to a private sector entity, a non-government organization (NGO), a local government, or to a local-level organization with farmers at its base. The most common form of IMT involves the shifting of management responsibility from a centralized government irrigation agency to a financially-autonomous local-level non-profit organization which is either controlled by the water users of the irrigation system or in which water users have a substantial voice in the control process.

The changes in management reported in the five primary case studies on which this paper is based can all be considered forms of IMT. However, there is an important difference between the organizational form of the recipient organization employed in the Philippines, on the one hand, and the form employed in the other four cases (Mexico, Turkey, Colombia, and Argentina). In the Philippines, the primary management unit employed is "community-based" and results from an intensive grassroots organizational campaign involving hired community organizers. This primary management unit is fairly small (less than 100 hectares), relies primarily on voluntary labour in carrying out its functions, and the most important relationships among members of the unit are social. In the other four cases, the organizational form of the irrigation systems can be termed an Irrigation Districts (ID). Irrigation Districts are typically larger (several thousand hectares), rely principally on paid employees to perform its functions, and link farmer-members together mainly through ties of economic self-interest.

Issues are analysed in this paper using a set of three basic categories:

(1) the processes used to introduce programs of IMT;

(2) the impacts of the introduction of a program of IMT; and

(3) second generation problems and possible solutions.

The term "second generation" requires some explanation. Transferring substantial management authority to a locally-based organization is a complicated undertaking and may involve changes in national policy, regulations, and organizational structures; creation of new organizations at the local level; transference of equipment ownership; and changes in personnel, in addition to the shifting of management functions to the new managers. Any undertaking this complex, in addition to solving problems, will almost certainly create new problems which did not exist before or were not previously evident. An example might be inadequate technical capability of new irrigation field personnel. These problems are here termed "second generation problems." In addition, there may be situations, such as low agricultural productivity, which were present prior to the transfer, but which were not acute problems when irrigation fees were low or non-existent. For our purposes, these are also included in the category of second generation problems. Some second generation problems may be a result of faulty processes used to introduce the new management system. Some may be a result of conscious choices during implementation to defer consideration of certain potential problems in the interests of accelerating program coverage. Others may be virtually unavoidable, though the ability to anticipate major problems in advance should allow corrective measures to be put into place earlier than would be otherwise possible. In this paper, problems are analysed from several different perspectives — that of the water user, the irrigation association, the irrigation agency, and the government. In terms of these perspectives, a change, such as increased irrigation service fees, may constitute a problem for water users, but a benefit for the irrigation association and perhaps the government, if it reduces government subsidies.

The term "impacts" also requires some explanation. In general, impacts can be either positive or negative. When they are positive, they are "benefits," or "positive benefits." When they are negative, they are similar to what are called second generation problems here. In the paper, negative impacts will be noted in the discussion of impacts, but will be discussed in more detail in the following section on second generation problems.

Process of Introducing New Forms of PIM

Background Conditions

Considering the case studies as well as PIM programs in other countries, key background conditions leading to the turnover process include:

- national budgetary crisis
- top level political will to place irrigated agriculture on a sound economic footing
- progressive deterioration of irrigation infrastructure due to deferred maintenance.

Only in one case (Colombia) of those reported at the workshop was it the farmers who initiated the process. At the irrigation systems level, another set of conditions that must be taken into consideration include:

- physical condition of hydraulic network;
- the social, political and economic conditions; and
- water availability.

Political will at the highest level of the government was a main component in Mexico's IMT program. In 1989, with a new administration in office, comprehensive water management was recognized as a top priority issue, the National Water Commission (CNA) was created and a national policy on privatization took off. In Colombia it was the National Planning Department who in 1991 submitted the Land Reclamation Program 1991-2000.

In Turkey, a budgetary crisis led to a squeeze on financial allocations to the state water resources agency (DSI) provided the initial impetus. In Argentina, a program to modernize the entire economy, which began in 1990 with the privatization of large electricity utilities, led to the turnover of water management to the provinces. In the Philippines the NIA embarked on an ambitious program in 1974 to increase rice production. A provision of this policy was for subsidies for the O&M of systems to be gradually phased out over a five year period at which time NIA would be directly dependent on collections from farmers for its O&M expenses.

The hydraulic infrastructure should be in fair condition and an affordable and reliable water supply should be available most of the time. Being in fair conditions means the hydraulic infrastructure is in operating conditions capable of delivering water to farms in sufficient amount to satisfy crop needs and in a timely manner. Surface drainage of surplus water and salinity should not be limiting factors. If these basic conditions are not satisfied, then a rehabilitation plan should be considered.

In Mexico, although the irrigation districts had suffered considerable deferred maintenance, the IMT program took off quickly because all of the systems were performing satisfactorily at the outset, with water conveyance efficiencies at main and secondary canal levels on the order of 60%.

The readiness of users to assume management responsibilities has to do with political and social factors. For example, it seems clear water users from Saldana and Coello in Colombia were "ready" in 1976 when they asked their government to turn over the administration of their districts to them. Due to its political background and social context it is apparent that Argentinean farmers were also ready for the change. In Turkey, farmers were used to a tradition of strong central government and many, especially the early adopters, were market oriented producers. This made them somewhat receptive to government declarations of the necessity of the transfer of managerial and financial responsibility. In most districts of Mexico, however, this was not the case. In the first place there was the land tenure issue. After the 1910 Revolution land was divided between two types of small farmers, *ejidatarios*, who worked small plots of land held communally, and small landowners, some of whom belonged to the *hacendados* or landlords' elite. Second, there were numerous voracious, government controlled *ejido* leaders who, when consulted about the coming change, voted against it suspecting that the change would mean a reduction of their present status. Promotional aspects of IMT in Mexico country were vital in order to overcome these difficulties.

National Policies

National policies for IMT implementation vary both in regard to objectives and scope. Under the Philippines Water Code of 1976, the appropriation of waters by an irrigator association has priority over requests of individuals. The government then helped farmers organize themselves into irrigator associations which enter into various types of contracts with the NIA to handle O&M of the system.

In Turkey the General Directorate of State Hydraulics (DSI) is the main executive agency of the Government for the country's overall water resources planning, execution and operation. It was established in 1954 and is part of the Ministry of Energy and Natural Resources. Since the early 1960s, DSI has had a program to transfer O&M responsibility for secondary and tertiary distribution networks to IAs. Under the program IAs entered into contracts with DSI to take administrative responsibility for tasks such as collecting and submitting farmer water demand application forms to DSI, managing water distribution below the secondary canal, and cleaning and minor repair of canals and other small hydraulic structures. Although existing municipality law appears to be providing a workable initial basis for the formation of IAs, their further development and evolution may require a law specifically for IAs.

In Argentina the country is a federal entity divided into 23 provinces, which are autonomous in all aspects related to water (rights, granting, duration, taxation, etc.). IMT programs began in 1990 as a response to pressure from the national government to reduce bureaucracy and render public administration more efficient.

In Mexico the 1989 presidential decree that created CNA also granted the agency the responsibility to: (a) define the country's water policies and (b) allocate water to users through licenses and permits. The new policy:

- Created autonomous and self-financing and water utilities to provide water services in cities and in irrigation districts,
- Encouraged water reuse and water quality conservation, and
- Promoted a new water culture based on the efficient use of the resource.

In Colombia the policy of IMT is part of a larger shift in Agricultural Policy towards minimizing state subsidies and regulation. However the state continues to play a major role in land reclamation, and rehabilitation and expansion of irrigation areas.

Irrigation Service Fees

In the case of Mexico, the IMT program was part of a series of changes in the economy including reductions in subsidies for agricultural credit and inputs, elimination of guaranteed support prices for the major agricultural crops, and increases in energy and fuel prices. Transfer of O&M responsibility for the irrigation districts, leading to users paying the real cost of irrigation water, was seen as just another step in the liberalization of the economy. In a similar fashion, in the case of the Philippines, a 1974 Presidential Decree authorized the National Irrigation Agency (NIA) to delegate partial or full management of irrigation systems to duly organized associations. Under this decree, NIA was allowed to keep all irrigation service fees, with government subsidies for O&M expenses gradually phased out over 5-years. Thus, at the end of this period NIA was to be directly dependent on ISF collections from farmers for O&M expenses.

The financial aspects of IMT in Turkey are similar to Mexico in that the policy is designed to shift the burden of O&M costs from the government to the users. However, the government continues to subsidize maintenance, which is not the case in Mexico and the Philippines. Colombia has also shifted the financial burden of O&M to the users, while irrigation schemes in Argentina are under joint management with fee collection by both the government and the IAs.

Water Laws

National water laws that clearly specify the rights of the IAs and the individual users appear to be an important factor in successful management transfer. Without such rights, the IAs are extremely vulnerable to increased demands from other more powerful users, such as industry and municipalities. Beginning in 1976, the Philippines, attempted to develop a water rights register of all water rights in the country, including specifying in volumetric form the water rights of the national and communal irrigation systems. As part of the registration process all IAs must register their water rights. Once legally registered, water rights cannot be withdrawn except for failure to use them as stipulated in the law.

This system can be contrasted to that of Mexico where the IAs are given a limited concession to use the irrigation infrastructure and the associated water supply, but do not have a clearly specified legal right to a volumetric supply. With domestic supply having priority, the IAs are not ensured a constant water supply over time. The concessions are also for a fixed time frame (20-30 years) after which they can theoretically be reassigned to another user. Since none of the concessions have expired to date, it is unclear exactly what process will be used to determine granting of a second concession.

The country where IAs appear to be most vulnerable to changing water demands is Turkey, where water is controlled by DSI and the individual associations have no water rights. This system works in areas where there is little competition for water but leaves the associations extremely vulnerable in areas where municipal & industrial use is expanding rapidly. Colombia is somewhere in between these two situations but still is highly dependent upon the national irrigation agency, as the country has yet to establish a legal water rights system.

The Mode of Implementing PIM

The process of implementing a change to a more participatory type management system varies widely from the bottom-up approach used in the Philippines to the more top-down approach used in Turkey and Mexico. In promoting the PIM process, Colombia has invested much less time and effort than the other countries. As a consequence, the process of transfer was very quick but there have been second generation problems.

Users have been less well informed and have been uncertain about their rights with respect to ownership and changes in management practices. As a number of the irrigation systems in Colombia are based on river pumping, some of the more recent cases of transfer have involved additional negotiations with respect to energy subsidies. Argentina has used information meetings and word-of-mouth to make a rapid transition to PIM, while Mexico and Turkey have used more structured informational meetings with users. These two countries also invested heavily in training their own staff. In Mexico, audio-visual materials prepared by IMTA and outside firms were used to persuade users that IMT was a positive change. The Philippines has used the slowest process with intensive use of institutional development officers and farmer organizers (FOs) to serve as catalysts. These organizers lived in the villages and organized exchanges between NIA and the users. However, NIA has now realized that too much dependence upon FOs has slowed the process and has now reduced significantly the number of FOs.

In terms of transferring responsibility, most of the Colombia IAs appear to have had very little say in the process. However, this has changed somewhat as more recently the IAs have been negotiating energy subsidies and insisting upon rehabilitation prior to transfer.

In the Philippines there has been more dialogue through the organizers but given that only limited transfer of responsibility has taken place, this approach has not achieved rapid transfer of responsibility to the users. In contrast, both in Turkey and in Mexico there have been very active negotiations concerning transfer terms. As a result the IAs have been able to exert some power and develop a partnership with the agency based on a meeting of equals. In a number of cases they have also been able to ensure that critical investments were made prior to transfer. As a key stage of the transfer process, Mexico, Colombia and Turkey have adopted a phase of shared management between the agency and the IAs during the transition. The duration of this shared management phase varies by country and by district but is usually between 6-12 months. In contrast, in Argentina the systems were transferred much more quickly, while in the Philippines there has been a very gradual shift in responsibility, in some cases over a period of many years.

Type/Nature of IA

The type and nature of the IAs are directly related to the structure of the economy as well as the type of irrigation found in the countries. In Mexico, Japan, Taiwan, Turkey, Argentina and Colombia, the IAs are larger (2,000-50,000 ha) and are organized more along the lines of commercial entities, reflecting the more commercial nature of the irrigated sector. Agriculture is developed on a cash basis and many of the staff are hired professionals paid in cash from the ISF. Given their large size, the IAs can afford to purchase and maintain their own transport and maintenance equipment.

In contrast, IAs in the Philippines are very small (100-300 ha) and are often organized based on the village structure. Most of the labour is voluntary labour provided by the users, and very few, if any, of the irrigation staff are hired professionals. Given the small size of the IAs there are diseconomies of scale and hence the organizations cannot afford to own specialized maintenance equipment. In the Philippines, irrigation service fees are usually paid in grain and therefore are very awkward to store and transport and typically result in 10-15% losses due to damage during storage and transport. In Argentina, IAs are public NGOs with full legal authority, including the power to tax. The IAs in the other countries have a limited power to establish and collect ISFs but do not have any other local taxation powers and therefore the majority of their income is from ISFs.

Impacts and Benefits of Transfer

Implementing a program of management transfer is a complicated undertaking which involves incurring costs and affecting the lives and livelihoods of many people. It is thus not desirable to enter into such a program unless the benefits of the changes are positive and significant. Impacts, of course, may be either positive or negative, and they can be either qualitative or quantitative. And because the change in management patterns will usually occur simultaneously with other changes in physical, economic, and social conditions, it may be difficult to separate the effects caused by management changes from those caused by other factors. The nature of the impacts which occur will be shaped by the social, political, and economic characteristics of the countries involved. Impacts are also conditioned by the perspective from which they are viewed.

Important perspectives in this case include those of water users, the associations they have already created, the irrigation agency and the national or state government under whose overall control these systems operate. What is positive from one perspective might be negative for another. Judgment is thus required in evaluating the overall impact of a program, and the tradeoffs in positive and negative benefits among the various groups affected.

Legal and Organizational Impacts

Some of the countries first build on a new national water Law before moving into a transfer program (Mexico, the Philippines and Colombia). Others move into transfer supported only by old national laws (Turkey) or local laws (Argentina). There has been a lot of discussion on the issue of water rights from the perspective of ownership versus concession. In most of the countries, all water belongs to the governments, either at federal or provincial levels. Irrigation Districts were usually created based on constitutional mandates that clearly defined the system's physical limits and thus the water right implicit to it. In some countries such as Argentina, however, the right of individual appropriation takes legal precedence.

After turnover, irrigation associations (IAs) in some countries receive the concessions which give them access to collective water rights. This does not change the existing system of water rights and priorities. In other countries, systems of water rights allocation and protection are virtually non-existent, which can lead to serious second-generation problems. In most of the case study countries, agricultural uses are subordinate to municipal uses, which include domestic and industrial use.

Sense of Ownership

The formation of IAs generally creates a stronger sense of ownership of the system on the part of users. In Japan, the Philippines, Colombia, and Argentina (Mendoza) this has been the case for many years. In Mexico and Turkey this change has taken place more recently and has been an exhilarating sense of pride for some users. Even though actual ownership of system facilities remains with the government. In Colombia some consideration is being given to transferring actual facilities ownership to IAs, an idea supported by some INAT staff and IA officers. This idea has also been discussed in the Philippines. There are a number of problems with this step, including loss of the right of eminent domain which is held by governments only and allows land condemnation for system expansion, and increased exposure to liability for accident and injury on the part of the IAs. Such a move may also weaken an IA's claim on public assistance for rehabilitation and emergency repairs.

A sense of ownership is also enhanced because users now have a greater voice in selecting the governance of their system, normally by electing boards of directors through the IA general assembly. With leaders elected in a democratic manner, needs can be expressed more readily. The opposite situation is presently occurring in Taiwan, where the users are turning systems back to the government. This is a result of the declining importance of agriculture in the Taiwanese economy and the high costs of system O&M which cannot be sustained by many IAs.

Transparency

All of the countries report that their transferred irrigation districts are now managed with more transparency, meaning by this that major decisions are exposed to the association's members or their representatives through board of directors and general assembly meetings. In Mexico, a representative assembly, formed by delegates selected by the general assembly takes routine decisions without summoning the general assembly, which may be composed of hundreds or even thousands of users. In Turkey a similar body made up of 30 to 70 local government officials and farmers also serves as a general assembly.

Transparency may be even more important in financial accounting than in organizational politics. Before transfer, the government collected fees and was generally supposed to invest the fees back into the system where they had been collected. However, due to bureaucratic procedures collected fees were not necessarily applied in the same systems where they originated. This resulted in complaints by users, and the agency was seldom forthcoming with information on financial accounts relative to particular systems. With associations collecting and managing funds, and with accounting done on the basis of the unit managed by the IA, users can get a better understanding of how their money is being used. However, strict supervision and auditing of water fee collection and expenditures by an outside party is still necessary to counter the possibility of money being diverted for unauthorized purposes.

Where the transferred units do not include an entire hydrologic unit, it may be more difficult for users to develop a sense of the disposition of the fees collected from them due to sub-division of responsibilities among a number of IAs. In this case, accounts should separate out the portion of fee collections that are earmarked for O&M within the IA unit, and the portion allocated to higher level O&M. An alternative might be for the IAs to merge into a larger unit. However there a number of other factors to consider before advocating such a step.

Water Fees Increase

Usually one of the immediate effects of IMT is an increase in water fees at the irrigation district level. In some countries the increase can be very dramatic, on the order of 100% to 200% or more (Mexico, Turkey). However to make a valid comparison, fee rates must be put in constant value terms before being compared. Irrigation service costs are generally considered appropriate if they constitute 5% to 8% of total production costs. In most countries, even after transfer, they tend to fall into this range. In the case of small holding sizes and lower value crops, however, the gross margin retained by farm households is also a relevant factor, and where this is already small, a doubling or tripling of ISF can create hardships for small farmers and their households. A proposed increment in ISF may also have a negative psychological impact on water users that have come to expect heavily subsidized irrigation service. This kind of attitude may be gradually changed with promotion activities before transfer takes place and, of course, with efficient financial management and better service once the association is formed.

ISF collection rates are above 50% in all five countries currently implementing IMT activities, and are above 70% in four of the five. However, in looking at the share of O&M costs covered by ISF collections, the picture is a bit different. For example, before transfer in Mexico, the ISF collection rate was around 100%, but this amount covered only 25% of O&M costs. After transfer, the ISF collection rate is still 100% in transferred districts, but fees now cover close to 90% of O&M expenses. In order for transferred schemes to be financially self-reliant, they must achieve high rates of fee collection as well as having an ISF that covers O&M costs.

Subsidies

From the government's perspective, an increase in O&M cost recovery normally means a reduction of subsidies. By subsidy we define here any kind of costs associated with the provision of irrigation services that the farmers should be paying instead of the government. Expenditures within an ID are usually related to:

- O&M costs plus overheads,
- acquisition of premises,
- purchasing machinery and equipment,
- rehabilitation and/or modernization,
- technical assistance and training, and
- emergencies.

O&M plus overheads refer to expenses related to the daily administration of the system from the head works down through the tertiaries. These expenses have to be paid in full, either by the government, the users, or a combination of both if a system is to be sustainable. The sharing of costs between the government and IAs ranges from 40/60 to nearly 0/100, as is the case with Mexico.

Acquisition of premises refers to the land and buildings IAs use as they provide service to farmers. In Mexico CNA initially provided premises to IAs as part of the concessions. After six years of turnover though, some associations have purchased their own offices, which they often refer to with pride.

Providing machinery and equipment to IAs has been a key component of transfer programs in Colombia and Mexico. In other countries, like Turkey and the Philippines, machines are still under the control of the irrigation agency. In Mexico equipment provision was heavily subsidized by the government to encourage farmers to participate in the IMT program. Many of the machines given in concession to IAs were old and the government had to subsidize repairs. Other machines were purchased under a World Bank loan. As lowering of maintenance costs was a concern for both government and users, emphasis was put on the purchase of light machines to substitute for the heavy construction equipment which had traditionally been used in the ID by CNA.

Rehabilitation and/or modernization of hydraulic infrastructure is also a key issue for all countries involved in IMT programs. Although farmers usually ask for investments of this kind to be made before turnover takes place, governments are often unwilling to supply the necessary resources to fulfil these requests. Turkey is an exception to this rule, where rehabilitation was not an integral part of the IMT program. In some cases countries use international loan funds for this purpose, enabling the government to negotiate with farmers the amount of needed investment and/or the financial share that users will need to contribute. Costs are often shared between the government and IAs on a 50/50 basis (e.g., Mexico).

Support for technical assistance and training may also be subsidized through IMT programs. In Mexico, technical training for both IAs and government irrigation district staff, has been provided by CNA with assistance from the Mexican Institute for Water Technology (IMTA). More recently, starting in 1995, some IAs have signed private agreements with IMTA in which technical assistance is being provided on specific projects with a limited subsidy from the CNA. In the Philippines, promotional and technical support to farmers is regularly provided by the government through various types of community organizers. In Turkey, extensive training for agency field staff and farmer leaders was provided by the irrigation agency. Yet, in all countries the training needs for PIM programs by far exceeds current investments in training.

Subsidies are also normally available in the case of emergencies situations. Although little specific information on this is available, all of the countries recognize the need of governmental intervention when such cases do arise, e.g., hurricanes in Mexico or typhoons in the Philippines. Water scarcity also creates emergency situations for transferred IDs. For example, during the 1995-96 irrigation season, an extraordinary drought took place in northern Mexico. Due to the lack of water, IAs did not have water to deliver to farmers; consequently, and collection of fees dropped nearly to zero. For many of the IAs the government had to step in with 100% subsidy programs to avoid a collapse of the associations. A few IAs that were only partially affected by the drought tackled the problem by hastily imposing a compensation fee on their farmers to implement a water reuse program.

Impacts on Operational Procedures

It is expected that a shift to participatory local management will improve the effectiveness and efficiency of water delivery as well as the quality of maintenance. The case studies provided some details to document these impacts but there is still a need to obtain better information about the operational impacts of IMT.

Improved Maintenance Programs

One of the immediate consequences of IMT programs is that irrigation systems have improved maintenance programs. (Colombia, as recently reported, seems to be an exception). A recent study carried out in 1994 in Mexico by the Colegio de Post-Graduados reported that 84% of the sample users believed maintenance had improved substantially since IAs have been in charge of O&M. In some areas not only have IAs completely eliminated the problem of deferred maintenance, but are also are now putting surplus money from O&M activities into programs for the modernization of hydraulic infrastructure.

Improved Services

Although few specific studies on quality of services after IMT exist, the general impression is that, after turnover, services have substantially improved in regard to timeliness, reliability, and equity of water distribution. There are many reasons to believe this. In the first place, after turnover users usually have better access to the irrigation service provider. Compared with the relationship with the agency that existed before, the farmer-association relationship is much better in terms of distance, personal contact and feedback on complaints. Water conflicts among farmers also tend to be minimized because farmers usually search for solutions to their problems at the

IA level which is physically closer to them. There is also a degree of self-control prevalent, as the users know each other very well and are therefore able to regulate behaviour among themselves.

In a 1994 study of IAs in Mexico, 84% of the users indicated that water distribution had improved since turnover, 79% said they were receiving enough water, 79% were also receiving water in time, and 64% indicated water was being allocated in the appropriate amount. However, water conflicts do arise between the IDs and IAs, especially when water is scarce. Recent lessons from the Mexican drought indicate that, under extreme competition for water, IAs were overwhelmed by operational problems. This required prompt government intervention to keep the conflicts from getting out of control and becoming an explosive political issue.

Agency Re-organization

IMT programs can have a significant impact on agency organization. A sharp reduction in field personnel and O&M staff usually takes place. A change in the role of the agency with regard to water management in the systems follows.

Staffing: Once the government's employees perceive the possibility of losing their jobs as a consequence of IMT implementation, they may attempt to block the turnover process. Clear legal arrangements between the parties are needed in order to overcome difficulties. Unfortunately, as the number of people involved in the process may be quite large. In Mexico, 5,000 out of 7,000 government personnel were to be released from the IDs because of turnover, while the Philippines has reduced government agency staff from 19,353 to 10,368.

Changing the role of the government irrigation agency is a natural outgrowth of IMT. The change requires not only willingness from the part of high-level officials but also training for them on specific issues. One is simply convincing governments officials that there have been substantial changes in the way the irrigation sector is structures. For instance, in some countries the agency keeps asking from IAs the same types of information as if management had never been transferred. In Mexico, after six years of turnover, changes in the governmental sector are still incomplete.

Once the turnover thoroughly takes place in a system, the agency's new role should be more oriented towards supervision, guidance, monitoring and regulation of water rights, plus selected technical assistance support to IAs. Management responsibilities of the system should be in the hands of the users, with the possible exception of head works which when considered of strategic importance or technically sophisticated, should remain under agency's direct control and administration.

Second Generation Problems and Solutions

As noted in the introduction, problems and solutions in a particular location will differ, depending on the perspective of the observer. The discussion below addresses problems from the perspectives of (a) the irrigation association, (b) farmers, (c) the irrigation agency, and (d) the government. Solving these problems can involve a variety of steps including revision of laws or implementing regulations and changes in organizational structure, organizational rules and processes, and funding mechanisms.

In addition, associations will require supporting services. Support services are services which come from outside the association itself but which are necessary for it to carry out its mission. They include such things as financing for equipment purchases, legal advice, computer programming assistance, and financial auditing services. Such services may be difficult for an association to generate internally (e.g., financing for heavy equipment) or be used only infrequently, (e.g., specialized maintenance equipment) and hence be too expensive for an association to maintain on a full-time basis.

Support services can be procured by associations from a variety of sources—private firms, public agencies, universities and institutes, non-government organizations, and regional or national federations of associations. In the past, it has usually been assumed that any such services must come from government agencies. Today it is recognized that higher quality and less expensive services may often be obtained from other sources, and that the government should generally serve as only one of several alternatives, rather than the sole source.

Irrigation Associations

Insecurity of water rights was identified as the most serious second generation problem affecting irrigation associations in the five case study countries. Water rights which are often absent, poorly-defined, or insecure, can (a) inhibit investment in new system facilities or rehabilitation, (b) encourage short-term thinking and behaviour on the part of association managers and farmers, (c) result in heavy expenditures in legal costs to defend a poorly-defined water right, and, ultimately, (d) lead to a reduction in water supply and even system collapse.

An effective water right should provide security to the association, but must, at the same time, be adaptable so that water can be diverted to other more productive or higher priority uses as economic and demographic conditions change. In this event, there must be a provision for appropriate compensation to those who are giving up the water right by those who gain it. In Mexico, for example, an association's irrigation water right is always subservient to present and future municipal demands. This creates considerable insecurity for associations which share water sources with growing municipalities and violates both the principles of security and that of just compensation. An effective water right should be specified in both quantitative and qualitative terms. Water quality degradation by upstream effluent discharges, as from a factory or an inadequate municipal sewage treatment plant, can render the water unusable by downstream irrigators. It can also make the water suitable only for lower value crops, since biologically or chemically contaminated water may not be permitted for production of higher value fresh fruits and vegetables. This will become an increasingly serious problem as water reuse increases in response to growing demand from all sectors.

Establishing a water rights system where it is lacking, as in Turkey, or clarifying water rights where they are weak, ineffective, or inequitable, as in Mexico, will usually require action from the national legislative body or from top level executive leadership, or both. It is thus extremely important for water user associations to have adequate representation of their interests when these issues are taken up. Two different types of support services are identified as being crucial for associations in attempting to establish or clarify their water rights. The first of these is legal advice and representation when the association faces challenges to its rights. Such representation is best secured from private law firms, if available, since services secured from government sources may be of lower quality, and may be subject to political pressures which would compromise their objectivity.

Legal representation is also important during the formative stage of an association, when negotiations with the government irrigation agency will establish the contract or concession which will control the relationship of the association and the government. Unfortunately, associations which are just forming may be unaware of the importance of high quality legal advice at this stage or may be unable to afford it. A national federation of associations can play a valuable role as a source of legal advice and assistance to newly forming associations.

The other type of support service required by associations is lobbying on their behalf in government policy making councils. Since other interests, such as municipalities and industrial water users, are usually larger than individual associations and likely to be more powerful politically, it is important for associations to establish regional or national federations representing many associations and a large number of farmers. This will give them political influence with which to counter the lower of competing interests. A federation would also be to represent irrigation associations on the board of directors of the national irrigation agency as is currently the case in Colombia.

Financial shortfalls comprise another high priority second generation problem affecting associations. A central feature of the PIM programs undertaken in all of the case study countries is financial autonomy. Financial autonomy is the condition where an organization generates all of the revenue it needs to support itself and to perform it's primary functions.

It implies that the association is not directly subsidized by the government, or that if it is subsidized, that the subsidy is a fixed amount which does not vary according to the condition of the association's balance sheet. The principal source of revenue for most associations is irrigation service fee (ISF) collections. Financial shortfalls affect a number of associations in Colombia and Argentina have been a concern in Mexico since the 1995 economic crisis.

Financial shortfalls are a function of several factors, including ISF rates, ISF collection effectiveness, the contribution of other sources of revenue, and expenditure patterns. One important factor is the structure of the ISF. Fees can be levied on flat rate or volumetric bases. A recommended structure for fees is a two-part one comprising both fixed and volumetric components. The flat portion would constitute a "connection charge", a charge for simply being within the boundaries of the system's service area whether or not water was actually taken from the system. This would reimburse the association for expenses incurred in maintaining the physical and administrative capacity to deliver water to the farm. The absence of this component in the fee structure of Mexican associations has created severe problems during years when drought greatly reduced the available water supply to the system's water users. The other portion of the fee would be based on the volume of water actually delivered during a cropping season, or some proxy for this amount, such as area irrigated and number of irrigations given. This would cover the costs incurred by the association which are related to the amount of water given, and would serve to limit excessive demand for water. Revenue from ISF is also dependent on the percentage of the fees assessed which are actually collected, though associations in many countries do a reasonably good job in this regard. An exception is the Philippines where low collection rates have been a persistent problem for NIA.

Solving problems of revenue shortfalls that relate to fee levels and collection efficiency is largely an internal association responsibility. Outside assistance may be useful, in some cases, in estimating farmers ability to pay particular ISF levels, and in analysing management systems set up for collecting revenue. This is discussed further in a following section. Underlying difficulties in generating sufficient ISF revenue to sustain system operations, in many cases, is the low productivity of irrigated agriculture in system command areas. Low productivity can result from a large number of factors, but is often associated with small farm size, a subsistence orientation, production of low value crops such as grains, inappropriate agricultural policies, a poor natural resource base, and inadequate agricultural support services.

In such cases, a solution to the association's financial problems may be possible only if the underlying problems in the agricultural sector are addressed. If these problems cannot be solved, then the options are for the government (a) to have other agencies provide technical assistance to increase production, or (b) provide the association with special subsidies. As a last resort, the government may have to consider taking back the responsibility of system management and financing. However, as irrigation service fees typically constitute only 3 to 10 percent of total production costs, reducing them will generally not solve underlying problems of high agricultural production costs and low productivity.

Rehabilitation is identified as a third important second-generation problem for irrigation associations. All irrigation systems require periodic rehabilitation and modernization. While usually less expensive, in real terms, than the original construction, rehabilitation is a costly undertaking, and is usually beyond the financial and technical means of an association to undertake. Seldom is there a clear and consistent government policy on responsibility for rehabilitation. In the absence of such a policy, the tendency is for associations to defer needed rehabilitation in the hope that the government will step in and take responsibility for it. In this case, IAs usually underinvest in system improvements between rehabilitations. This tendency is reinforced by the government retaining ownership of system physical facilities, while transferring to associations only the use rights of the facilities.

Associations may thus regard the responsibility to rehabilitate those facilities as belonging to the government unless a different policy is clearly stated in the agreement between the government and the association. A related problem is that of cost sharing between the government and the association for rehabilitation. Because irrigated agriculture benefits people beyond the ranks of system irrigators, and because full coverage of rehabilitation costs is usually beyond the means of the irrigators themselves, a sharing of costs is appropriate. Responsibility for even a share of the costs involved will tend to counteract the tendency of an association to defer maintenance, as noted above. To cover its share of future rehabilitation costs, associations usually need to accumulate a capital replacement fund over a number of years and there needs to be a legal basis for establishing such a fund.

There should also be incentives for establishing and contributing to such a fund. Unfortunately, fiscal and monetary policies in many countries such as Turkey and Mexico, have led to high rates of inflation and low or negative real interest rates on savings, which acts as a powerful deterrent to fund accumulation. Governments may need to create special investment opportunities for associations which allow them to earn reasonable rates of return on accumulated funds. Likewise, there should be incentives for associations to make improvements in physical infrastructure. One way to do this is to establish a trust fund, perhaps with donor financing, from which associations could request funds to complement their own investment funds.

The matching ratio for such a funding facility should be established and made known in advance. A number of support services are required specifically to support system rehabilitation. These include (1) assessments of system facilities, (2) credit, and (3) design and construction assistance. Regular assessments of the condition of system facilities can be done jointly by the association and the government agency, as in Turkey, or may be contracted out to an engineering consulting firm acceptable to both the association and the agency. Such assessments can be used as a basis for annual maintenance planning, to suggest the need for selective improvements in system facilities, and for planning whole-system rehabilitation.

If an association is unable to accumulate its share of rehabilitation cost prior to rehabilitation and does not have the ability to assess a special charge on the members, it will need a source of credit. Credit can come from private banks, government banks or other lending facilities, or from insurance pools in the case of rehabilitation induced by natural disasters such as floods, typhoons, or volcanic eruptions. Such a credit facility could also be used as a source of financing for capital equipment needed for system maintenance.

Rehabilitation will also require external technical services for design and construction. Because of the sharing of costs, both the association and the government should be involved in decisionmaking relating to the selection of consultants and contractors and monitoring their performance. Advice and guidance to the association on handing these tasks might usefully be given by a federation of associations, since rehabilitation occurs only infrequently in any one association.

Lack of financial and administrative management expertise is the final topic identified as a high-priority second generation problem by workshop participants. There are several possible responses to this problem. One would be skill enhancement through staff training programs. Skills can also be enhanced by replacing less skilled people with more capable ones. Contracting out for specialized services is another important way of addressing management deficiencies in associations.

One extremely important step in improving the quality of association management is to increase the transparency of management processes. This has a number of positive effects. It can (a) reduce the potential for misappropriation of funds, (b) help insure that salary levels and benefits are realistic, (c) help insure that maintenance allocations are appropriately targeted, (d) reduce favouritism in making personnel appointments, and (e) improve responsiveness of association staff to users. A number of steps can be taken to increase transparency in association management. These include:

- regular external audits of financial accounts,
- use of standardized budgeting and accounting frameworks,
- wide dissemination of simplified budgets, plans, and financial statements,
- active involvement of the board of directors in forward planning, budgeting, and auditing, and
- broad representation from among users on the association board of directors.

There is broad scope for employing external support services to improve management of the association. Services that may be required include:

- advice on establishing and revising management systems and procedures,
- advice on establishing financial budgeting and accounting systems, including software,
- establishment of standard budgeting and accounting formats,
- standards and requirements for regular external audits, and
- management training.

These services can be obtained by the association from a variety of sources, including private firms, a national or regional federation of associations, NGOs, government agencies, and universities and training institutes. The government is the logical party to establish mandatory standards and requirements for external audits, but the audits themselves, could be done by a private firm of chartered accountants. Other services, such as management and accounting system advice could come from a variety of sources, with private sources being generally preferred.

One argument in favour of the provision of these services by government agencies will often be that they can be obtained at no or low cost. What makes this lower cost provision possible, however, will generally be implicit government subsidies to the service providers. A preferred alternative would be to provide the funds supporting these subsidies instead to the associations as grants to be used for obtaining management support services. This would allow the associations to contract for these services among alternative providers. The demand-driven competition thus induced would be a very healthy force acting to hold down service prices and improve quality of services delivered. Provision of such grants during the transition phase from government agency to association management could be a very useful institution-strengthening activity.

Irrigation Agencies

Dislocation of staff is the most prominent problem experienced by agencies following irrigation management transfer to users. This problem is typically dealt with in several ways.

First, O&M staff levels are reduced by attrition. When positions become vacant due to retirement or resignation, they are left vacant or filled by internal transfers rather than new hiring.

Second, financial incentives are often provided for early retirement of older staff. Third, existing staff are transferred to other positions which become vacant rather than filling them from outside the agency. In some countries such as China, where it is difficult to lay off staff, sideline enterprises have been created which can generate income for the irrigation district and cover the salary costs of the involved personnel. In the Philippines, this has taken the form of an irrigation consulting company, NIA Consult, which was created as a subsidiary of the National Irrigation Administration to provide irrigation advisory and design services within the Philippines and abroad.

In some cases, redundant agency staff may also be re-employed by the newly created associations which take up management responsibility for the schemes. Such employment should be at the discretion of the association. Loss of technical capacity in the national irrigation agency is a common problem faced by agencies which transfer management significant irrigated areas to associations and experience loss of technical staff. To address this problem, agencies can

- obtain specialized expertise from outside consulting firms as needed,
- increase salaries to attract and retain high-quality staff,
- provide in-service training opportunities for staff, and
- revise job descriptions to bring in new staff with the desired qualifications.

Defining a new role for the agency is another important challenge. With their operational responsibilities transferred to associations, the agencies need to design a new role to address emerging problems. Doing this requires discussion among staff within the agency and also at higher levels of the government, with broad participation by all involved parties. The aim should be to build broad consensus and political commitment for the new role. In some cases, changes in legislation may be required to enable the assumption of new responsibilities.

The new mandate should contain a clear definition of roles and responsibilities and should define skill requirements to carry out the new responsibilities. It should also contain a timetable for accomplishing the shift to the new mandate. Support services which could be useful in this process include:

- comprehensive diagnosis of the agency/association relationship and the associations' support needs,
- professional assistance with the agency's strategic planning process, and

- consulting services to design new management information systems for the agency.

In Colombia, the national agency, INAT, is employing professional consultants to help them define a new role for themselves under an Inter-American Development Bank credit.

Farmers

Second generation problems experienced by individual farmers relate mainly to the need to increase farm productivity to pay higher irrigation fees and to take advantage of possible improvements in irrigation service quality. Support services required may include:

- production credit,
- extension advice,
- new technologies,
- markets and market information,
- access to inputs, and
- post-harvest services.

Although government agencies are the traditional source of many of these services, in many countries, private or other organizations are playing an expanding role in supplying some or all of the services listed above. There is also the question of the potential role of the association itself in providing other agricultural services, in addition to irrigation service. As a general rule, the association should demonstrate competence in its core activity of irrigation management before considering such ancillary activities as providing other agricultural inputs.

Government

The principal second-generation problem for government, beyond those already identified for the irrigation agency, is the reduced control which it will have over irrigation activities at the system level, and a diminished ability to use irrigation as a tool to implement other national policies and priorities. An example might be the government's wish to promote cultivation of upland crops rather than rice during a particular season.

In the past it could work through the national irrigation agency to adjust water delivery schedules and volumes to try to achieve this end. Following transfer, this becomes more difficult. There are other tools, such as support prices and subsidies, to achieve the same ends, however, so that losing control of irrigation system management should not pose a significant problem for agricultural policymakers.

Summary and Conclusions

Experience is now available from a number of developing countries which have recently implemented Participatory Irrigation Management (PIM) programs and there is additional evidence from developed countries which transferred irrigation management functions to locally-based organizations many years ago. In the case of the developing countries, while the overall benefits of PIM have been positive, in some cases second generation problems have manifested themselves and, consequently, have tended to reduce the magnitude of the potential benefits. In the process of introducing PIM programs, political will at the highest level is a critical background condition for a rapid and sustained transfer program. A second important element is that the irrigation infrastructure be in fair condition so that it could deliver irrigation water as required. A suitable legal framework is also necessary for the sustainable functioning of the transferred systems. Lack of clear water rights has led to second generation problems including conflicts with municipal and industrial users as well as other irrigation organizations.

PIM is designed to shift the financial burden for irrigation service from the agency to the users. This aspect has to be made very clear when the process of transfer is introduced. Failure to address the financial side of system management is a primary cause of second generation problems. In general, the countries that have a clear policy on irrigation service fee rates and collection practices have sustainable water user associations. The type and nature of the associations are very dependent on the structure of the broader economy as well as the type of irrigation and the tradition of management prevailing in the country. Where economies are more developed and diversified and irrigation systems are large, associations have tended to evolve successfully. These associations are generally large and can hire their own staff and own specialized irrigation equipment. In contrast, in countries where economies are less developed, agriculture is more subsistence oriented, and irrigation systems are small, the associations tend to be small and more problematic in terms of management.

These changes in management structures and processes have had important impacts both positive and negative on four important target groups: farmers, the irrigation association, the irrigation agency, and the government. For example, while increased service fees have reduced the financial burden on the government and increase the sustainability of the IAs, they have added to the costs of production for the users. On the other hand, from the perspective of the farmers, transfer has resulted in a sense of ownership, reduction in conflict and improved maintenance.

Transfer has reduced the O&M staff of the irrigation agencies, and consequently the overall number of civil servants working in the irrigated agriculture sector. However, in a number of cases, this has also reduced governments' control over cropping patterns and over water resources more generally. This reduced government control has generally led to greater farmer satisfaction, more transparency in decisionmaking, and greater overall economic efficiency.

Changes in management responsibility have led to second generation problems in most countries, some of which are already affecting the involved parties while others loom as potential future problems. Insecurity of water rights was identified as the most serious second generation problem affecting transferred systems. The primary solution is to establish a secure legal right that has both quantitative and qualitative dimensions. Federations of associations can provide assistance and legal representation.

Financial shortfalls comprise another second generation problem. The principle source of revenue for the association is the irrigation service fee. A two-part fee consisting of a fixed connection charge and a volumetric charge can provide greater financial stability. Increasing the productivity of irrigated agriculture is also an important element in improving the financial health of associations. Outside assistance may be useful in analysing and improving management systems set up to assess and collect service fees.

Rehabilitation poses a number of second generation problems. In the absence of a clear and consistent policy on rehabilitation, maintenance is often deferred. There is a valid argument for developing a cost sharing formula where the government pays a share and the users pay the remaining share. If possible, the IAs should have a sinking fund for accumulating funds to cover their share of future rehabilitation costs. When this is insufficient, it is important to have an outside source of affordable credit. Other needed supporting services include assistance with maintenance assessment practices and technical design and construction services.

Effective financial and administrative management of the associations requires specialized staff training and increased transparency. Support services such as external audits, and standardized accounting packages can also contribute to better management of the associations.

Irrigation agencies also suffer from a number of second generation problems. Dislocation of staff, loss of technical capability, and the need to define a new role for the agency are all problems found in countries that are instituting PIM programs. In particular, the problem of what to do with excess staff is a difficulty faced by almost all agencies. Solutions include attrition, retirement incentives, creation of specialized consulting units, retraining and assistance with job placement, and transfer to other units. Along with the problem of staff displacement, agencies also face the problem of the loss of specialized skills. These may be replaced by outside private consultants but may also require the agency to provide specialized training for existing staff.

Second generation problems of farmers are related to the need to increase agricultural productivity, including the need to shift to higher value crops. Services such as credit, agricultural extension, market access, technical inputs and post-harvest assistance are all needed. In some cases these services will come from the government but increasingly from the private sector as well. Federations of IAs can also play an important role in the provision of such services.

A shift from public agency control to local participatory management is unlikely to happen without some second generation problems. Rapid institutional change will almost always require corrective measures to address unexpected problems. Countries with flexible policies and procedures will be able to address these problems, as they arise. This report has summarized a number of solutions countries have employed to address second generation problems associated with the shift to participatory irrigation management.

6

Irrigation Methods

With agriculture responsible for the largest water usage in the United States and with irrigation dams being the most common type of water supply dam, it is important to examine the way this industry uses water and how conservation methods can be used to increase efficiencies and thus possibly decrease the need for dams. In addition to some of the alternative diversion techniques (described above) to supply water for irrigation, the U.S. EPA has compiled water-saving irrigation practices into three categories:

- Field Practices
- Management Strategies
- System Modifications.

When these practices are combined with the alternative diversion strategies above, the need for a diversion dam for irrigation could be eliminated in some circumstances.

Land has been tilled and terraced to better capture water (Lynn Betts, USDA Natural Resource Conservation Service)

Field Practices

Field practices are techniques focused on keeping water in the field, distributing it more efficiently, or achieving better soil moisture retention. These techniques are typically less expensive than management strategies or system modifications. When traditional field practices fall short of expectations and the management strategies and systems modifications discussed below are out of reach, the field practices of dryland farming and land retirement are another avenue to explore. Examples of field practices include:

The chiseling of extremely compacted soils;

- Furrow diking to prevent runoff;
- Land levelling for a more even water distribution
- Dryland farming; and
- Land retirement.

Farmers can develop land management practices that will decrease the demand on water supplies. More than half of land used for agriculture is still irrigated via a gravityflow system. This system uses soil borders, furrows, or ditches in order to allow gravity to distribute water across fields. Gravity flow irrigation methods can result in up to 50 percent water loss due to evaporation, inefficiencies in water delivery to the crop-root zone and runoff at the end of the field. The traditional gravity-fed system can be improved upon with the use of laser levelling or micro irrigation, though evaporation still leads to water loss. Laser levelling involves grading and precisely levelling the soil to eliminate any variation in the gradient and reduce slope of the field. This helps control the flow of the water and allows for more uniform soil saturation. Another method of preventing runoff is furrow diking. Furrow diking is the practice of building small temporary dikes across furrows to conserve water for crop production, which may also aid in preventing erosion.

If the above land management practices are not decreasing water use enough and the system modifications described below are too cost prohibitive or not an appropriate technique for a particular crop, farmers can also consider converting to dryland farming, switching to less water intensive crops, or land retirement. Farmers practicing dryland farming in arid regions use a variety of techniques and land management practices to minimize water loss and erosion. These techniques include coordinating seeding to the ideal soil moisture content, choosing crops more suited for arid conditions, and fallowing. Fallowing refers to a number of practices used for well over a century, such as plowing a field in late fall or early spring to clear weeds and increase soil moisture. Initial plowing breaks up the land and allows the soil to absorb more water. It also eliminates moisture-sucking weeds and creates ridges in the land that limit runoff and better capture moisture from snow. Fallowing can also involve choosing not to plant a certain field for one or more growing seasons.

Land retirement refers to a common policy of permanently or temporarily suspending farming on a particular acreage of land in exchange for financial incentives. One of the best-known land retirement programs is the U.S. Department of Agriculture's Conservation Reserve Program (CRP). Through CRP, farmers are paid annual rent per acre and an additional sum for providing land cover. While CRP has typically been utilized to control the agricultural market and keep prices and quantities stable, the added value of conserving land and water resources has been given more consideration in determining compensation for land retirement since the late 1990s. This type of financial incentive is common among land retirement programs.

Advantages

Practices such as chiseling, furrow diking, and land levelling allow the land to absorb water more efficiently and results in less waste. It is also one of the most inexpensive methods of agricultural water conservation discussed in this report. Depending on the amount of land in need of irrigating and the alternative chosen, it might be possible to remove an irrigation diversion dam, particularly if used in combination with one of the alternative diversion methods described above. Dryland farming and land retirement, also discussed above, have the most to offer in terms of water savings, simply because they call for the use of little to no water, and the potential for dam removal.

Disadvantages

While chiseling, furrow diking, and land levelling help prevent runoff and allow the land to retain more water, they still do not address the over watering that results from gravity-fed irrigation. Also, dryland farming and land retirement practices can seem akin to suggesting that farmers go out of business. Discussions centring on these alternatives should take current use and compensation into consideration. Also, dryland farming and land retirement practices are rarely, if ever, applied to the large agribusinesses that now dominate the industry.

Costs

As discussed above, furrowing and other land levelling practices are the least expensive irrigation alternatives discussed in this report. Actual project costs will vary depending on amount of acreage, topography of the land, and the region or country in which the farm is located. According to the 1998 Farm and Ranch Irrigation Survey, capital expenditures in the United States for farm improvements were $643 million for irrigation equipment and machinery, $138 million for construction and deepening of wells, $190 million for permanent storage and distribution systems, and $83 million for land clearing and levelling. In order for dryland farming and land retirement to be feasible for farmers, it often must be accompanied by financial incentives like conservation easements, which involves the transfer of development and/or land use rights to a government agency or non-profit providing tax benefits or direct payment for retirement of the land.

Management Strategies

Management strategies allow the irrigator to monitor soil and water conditions to ensure water is delivered in the most efficient manner possible. By collecting this information, farmers can make informed decisions about scheduling, the appropriate amount of water for a particular crop, and any system upgrades that may be needed. The methods include:

- Measuring rainfall;
- Determining soil moisture;
- Checking pumping plant efficiency; and
- Scheduling irrigation.

Farmers have to rely on a number of factors to monitor soil moisture, including temperature and humidity, solar radiation, crop growth stage, mulch, soil texture, percentage of organic matter, and rooting depth. A variety of tools for monitoring soil moisture, such as Time Domain Reflectivity (TDR) probes or tensiometers, are also available to farmers.

The government of Queensland in Australia has done an effective job of compiling a fact sheet on a variety of irrigation scheduling tools, including the associated pros, cons, and costs of each. Ensuring that pumping plants are running at their most efficient also guarantees that water is being delivered to the plant and not wasted. Efficiency can be checked by examining the volume of water pumped, the lift, and the amount of energy used. A pump in need of repair or adjustment can not only waste water but also cost money.

Advantages

The management strategies described above allow for the correct amount of moisture to be delivered to the plant. When combined with system upgrades like the ones discussed below, farmers can maximize the amount of water savings and the efficiency of their land. While this is not an automatic replacement for a dam, there could be an opportunity for removal or the ability to delay construction a new barrier, depending on the size of the diversion.

Disadvantages

Monitoring the water needs of crops in the most efficient manner possible requires technological upgrades that require an initial outlay of capital. In addition to the cost of implementing these system upgrades, there may be training required to integrate new computer systems and other technologies.

Costs

Depending on extensiveness of the system, costs can vary significantly for the management strategies discussed above. For example, the average price of a tensiometer ranges from $120 to $200, with the average field requiring a minimum of four stations containing two tensiometers each, while acprobe system containing probes, training, and software can run as much as $9,120.

The Department of Natural Resources, Energy and Mines in Queensland (DNREM), Australia has put together a comprehensive fact sheet that provides cost estimates (in Australian dollars) for a wide range of irrigation scheduling tools.

A centre pivot irrigation system with drop tubes (Tim McCabe, USDA Natural Resource Conservation Service)

System Modifications

System modifications, often the most expensive of the three categories, require making changes to an existing irrigation system or replacing an existing system with a new one. Typical system modifications that allow for the most efficient delivery of water are:

- Add drop tubes to a centre pivot system
- Retrofitting a well with a smaller pump.

Replacement irrigation systems include:

- Installing drip irrigation, microsprinklers, or solid set systems; or
- Constructing a tailwater recovery system.

Many farms still use inefficient irrigation techniques (e.g., travelling gun, centre pivot) that apply more water than crops require. Modern irrigation technology, such as drip irrigation, micro sprinklers and solid set systems can deliver water much closer to the actual plant and achieve much greater water efficiency. These irrigation tools are the most efficient in terms of delivering water to crops. They use the latest technologies to determine the exact amount of water a crop needs in order to grow and delivers the water directly to the plant. However, they often prove most efficient when used with vegetable and fruit tree crops and less so with dense grain crops.

Advantages

Because of the considerable amount of water used in agriculture, improving efficiency in this sector offers an opportunity to achieve significant reductions in water use. By using the latest technology available to maximize the efficient use of water, the need for some water diversions and dams can be eliminated.

Disadvantages

Switching to more efficient irrigation technologies is cost prohibitive for many farmers. Even though federal and state incentives exist, they are often inadequate to address the scope of the problem.

Costs

As mentioned above, initial costs of the latest irrigation technology can be quite high. For example, drip irrigation systems can cost on average $1,000 per acre to install necessary pumps and filters and $150 per acre per year for drip tubing.

A study done by Kansas State University Agricultural Experiment Station in October 2001 compared the costs of centre pivot, flood and drip irrigation systems. While the drip irrigation systems are typically more expensive to install, farmers are able to recoup some costs with savings from reduced water use.

Irrigation Scheduling

Irrigation scheduling is the process used by irrigation system managers to determine the correct frequency and duration of watering.

The following factors may be taken into consideration:

- Precipitation rate of the irrigation equipment-how quickly the water is applied, often expressed in inches or mm per hour.
- Distribution uniformity of the irrigation system-how uniformly the water is applied, expressed as a percentage, the higher the number, the more uniform.

- Soil infiltration rate-how quickly the water is absorbed by the soil, the rate of which also decreases as the soil becomes wetter, also often expressed in inches or mm per hour.
- Slope (topography) of the land being irrigated as this affects how quickly runoff occurs, often expressed as a percentage, i.e. distance of fall divided by 100 units of horizontal distance (1 ft of fall per 100 ft (30 m) would be 1%).
- Soil available water capacity, expressed in units of water per unit of soil, i.e. inches of water per foot of soil.
- Effective rooting depth of the plants to be watered, which affects how much water can be stored in the soil and made available to the plants.
- Current watering requirements of the plant (which may be estimated by calculating evapotranspiration, or ET), often expressed in inches per day.
- Amount of time in which water or labour may be available for irrigation.
- Amount of allowable moisture stress which may be placed on the plant. For high value vegetable crops, this may mean no allowable stress, while for a lawn some stress would be allowable, since the goal would not be to maximize production, but merely to keep the lawn green and healthy.
- Timing to take advantage of projected rainfall
- Timing to take advantage of favourable utility rates
- Timing to avoid interfering with other activities such as sporting events, holidays, lawn maintenance, or crop harvesting.

The goal in irrigation scheduling is to apply enough water to fully wet the plant's root zone while minimizing over watering and then allow the soil to dry out in between waterings, to allow air to enter the soil and encourage root development, but not so much that the plant is stressed beyond what is allowable.

In recent years, more sophisticated irrigation controllers have been developed that receive ET input from either a single on-site weather station or from a network of stations and automatically adjust the irrigation schedule accordingly. When properly set up and maintained, these controllers do tend to conserve water over conventional human scheduling as the program is updated at least daily. Other devices helpful in irrigation scheduling are rain sensors, which automatically shut off an irrigation system when it rains, and soil moisture sensing devices such as capacitance sensors, tensiometers and gypsum blocks.

Choosing the most Advantageous Field for Drip Irrigation

Drip irrigation is gaining large amount of popularity due to the wonderful amount of benefits that is being provided by this irrigation technique. Nowadays there is more and more number of farmers who are implementing drip irrigation method which gives them the opportunity to automate the process of irrigation. In order to maximize the amount of benefits that are derived through drip irrigation it is necessary that the person possess adequate knowledge regarding the selection of proper field that will provide him best results. This article will be providing good deal of idea that will help in increasing the harvest of crops and also improve the quality of plants. The factor which influences the selection of field is technology, nature of crops, climatic conditions, labour inputs and many more.

Soil Selection in Fields

In drip irrigation there will constant spraying of water uniformly around the field. The soil type that is most suitable for drip irrigation is sandy soils. Surface irrigation cannot be applied for sandy soils since these soils have high penetration and very less storage capacity. Thus drip irrigation is the only best method that makes sure that proper level of water is supplied to plants without water wastage. It is also essential to note that clay or loamy soils should not be selected for fields that are being irrigated by drip irrigation since it will make soil slippery and water logged.

Climatic Conditions and Slope of Field

Drip irrigation can be greatly affected if the wind velocity in field region is on the higher side. When the wind velocity is higher then the throw of water will be disrupted and non uniform spread of water will be made. The effect of drip irrigation can be maximized in which the flow of wind is uniform and stable. Another important factor which influences field selection for drip irrigation is slope of field. This irrigation technique will provide more benefits if field is sloppy or non uniform since it will ensure proper path for flow of water. Best example for sloppy field is rice grown in terrace space. Surface irrigation cannot be applied in regions which contains slope.

Water Quality and Type of Crops

One of the important requirements for drip irrigation is quality of water which is present in field. It is essential to select field space in which water contains very less level of sediments or salt particles. Presence of higher level of sediments or salt particles will clog the sprinklers and decreases the efficiency of water spray. Since there is large level of investment which is being made for this irrigation technique only those fields which supports growth of higher value cash crops should be selected to ensure the farmer gets good returns for the money invested

Available Labour

Drip irrigation provides the possibility of reducing the labour requirement due to automation process. Fields where lower level of labours is available will be well benefited by adapting drip irrigation technique. It is essential to maintain sprinklers in good condition to maintain efficiency in operation.

Using Drip Farming in Winter

Agriculture has found drastic changes in past few years. The world has witnessed revolutionized approach in carrying out tasks in the field. The on-field agricultural practices have been greatly supported by inventions and discoveries that have hit the science world. Also usage of sophisticated tools and equipments has gathered substantial interests of farmers. This has empowered them to build up comprehensive farming with less effort generating huge dividends for the money invested. Importance of ambient conditions can be visualized with development of hybrid seasonal crops. The modern scientific aids to modern agriculture are also provided through researches carried out all over the world. Farming has now channelled its path towards thoughtful use of conditions with optimized usage of resources available. For instance, one can get huge throughput by planting right crops at right seasons. For instance during summers, one can engage cultivation of crops that favour higher temperature during their growth. This makes the overall process of farming refined totally conducive for people to march down to sustainable development.

Basic Irrigation Facilities

Irrigation plays a major role in the development of crops. Water can be utilized judiciously for the growth of the crop. There are various types of irrigation facilities available for the farmers present globally. Starting from ground water supply to supply from rivers. Researchers have also modularized the irrigation facilities to cater to ever-increasing demands. The use of sprinklers can be used to cover large area of field in short duration of time. This can also be by-passed by usage of overhead irrigation. Some of classical methods of irrigation include terracing and furrow irrigation. Furrow irrigation involves constant monitoring of farmers to regulate channeling of water to respective fields. Recent developments of science and technology have included drip irrigation to this long list.

Importance of Drip Irrigation

Drip irrigation involves the type of irrigation wherein the supply of water is confined to the root zones of each and every plant. This can greatly lessen the fritter of water. Also one can be sure of getting necessary supplied nutrition to the plant. Every other types of irrigation include surface evaporation of water. This can reduce moisture content that is required for the growth of plants. The pressure of water channel can send the water deep down into the earth at desirable spots near the root zone. Also by utilizing efficient computer module, one can monitor the flow of water and timing of watering to crops. This can be cashing for labour in many ways but the operating cost can fetch fruitful results for farmers. In recent years, sprinkler type of irrigation is replaced by unique type of drip irrigation referred as subsurface drip irrigation.

Usefulness of Drip Farming in Winter

Drip farming includes the laying down structural lines for channeling water to respective spots underneath the soil. This can also be aided by several hydraulic components such as pressure relief valves and hoses. These can have handsome impact from lowering of temperature during winters. The temperature changes can affect any physical element by changing the microscopic properties of materials used. Drip farming has major impact during the winter as it can get hoses and hydraulic accessories to get frozen with water that is capsulated within them. So to avoid mishaps, one can adopt popular winterizing technique. This involves expelling the water contained through special equipments that employs purging air to accomplish this act. Blown air would remove all contaminants and undesirable particles present in hydraulic lines. Electronic components associated with drip farming methods can be stored in secured place avoiding direct exposure to moisture. Regular checking is to be instituted for retaining cleaner fluid lines over the field area during winter season.

Drip Irrigation Fittings

Agriculture is a vast field. Every human has to consume food in order to live. These food items are got only from the hardship of a farmer. As the technology improves many modern methods have evolved to sophisticate all including the farmers. Most of these focused only on reduction of burden on the farmer and not on the improvement of farming. One step was taken to provide optimized resource utilization in the agriculture field. The new invention and the one which takes care of this optimizing step is the drip irrigation process. Many are not aware of the drip irrigation process. Some of them are stubborn and are not ready to go in for another process. They feel safe and secure with the conventional process.

Why we have to go in for Drip Irrigation?

As we all no water scarcity is the most faced problem around the globe. Not sufficient amount of water is present with many villages and their crops get spoiled due to it. So saving of water is a must for all of us. This can be easily achieved by the use of drip irrigation process instead of the conventional process. The wastage of water can be drastically reduced when drip irrigation is followed. The working principle of drip irrigation process is very simple and uses small openings in the hose to water the plant. The holes are so small that only drops of water emerge from these holes. So the water usage rate is so minimum and this is even smaller than the water intake rate of the plants. So the wastage of water is minimal and the utilization of water is optimum. The type of hoses used for this process is the soaker hose. This hose has minute holes of 1/6 to 1/4 inch diameter. And this is also has unique capability to work both on the surface and under the surface. The most important thing that is to be mentioned here is about the fittings.

Fittings in the Drip Irrigation Process

Water leakage and water loss in the fitting is the major loss in the case of any hose usage. The wear and tear is mostly felt at the fitting region. This can be reduced drastically if the special fittings are provided. The fittings used in the drip irrigation process is one of its kind. The fittings are more precisely manufactured and they form a tight fit. The clearance in between the threads are so minimum. The material these fittings are made up of is the brass. In addition to it hoses also posses this capability to fit perfectly with any pipes.

Drip Irrigation Timer

Some of us would have come across the word drip irrigation. We all know irrigation is a process of providing water to the plants, but we don't know what is drip irrigation. Drip irrigation is nothing but, providing water to the specific plant at specific place at specific time. The conventional process of irrigation satisfies these thing, then why is that we should go in for a new process of irrigation. The answer to this question is very simple.

Why Drip Irrigation is Needed ?

The need for drip irrigation is due to the difference or advantage it has over the conventional process. The drip irrigation saves water while the others waste them. Water is a big problem faced by many of the countries. So saving water is very essential to all of mankind. The process of drip irrigation is very simple and its application is also very easy. The working principle involved is that the plants are provided water with minute holes in the hoses. These hoses are known as soaker hoses. The soaker hose works by the principle of capillary outlets. There are very small holes in the range of 1/6 to 1/4 of an inch. These holes form the outlet or source of water supply to the plants. The other great advantage in these hoses are that, it can fit with any type of fittings and pipes. This advantage is got by the hose due to the fact that it poses brass fittings at the ends. The higher end hoses even poses greater advantages than the lower end hoses. They are manufactured precisely to meet the standards. The hose can be placed above the soil or as in some cases buried in the soil. This is how people grow plants in dry places. They are highly advantageous and capable of providing plants in any soil conditions. The other thing is that, it saves water by reducing the sprinkling and showering.

Drip Irrigation Timing

The other important thing that is so far not mentioned about the drip irrigation is that it can be integrated with mechatronics system. The water is provided to the plants based upon the actuation of valves. The valve opening and closing time can be automated using mechatronics system. So the water is provided to the plants at proper intervals and proper amount of water can be given to the plants. The advantage of the drip irrigation is doubled on application of the timing system in the process.

Drip Irrigation Soaker Hose

Irrigation is a common terminology to all of us. But the real meaning of it, is not known by many. Irrigation is a process of watering the plants artificially. This is done because, sufficient amount of water may not be present in that area or the season may not be correct for the cultivation. There are many artificial ways of irrigation possible.

The normal way of irrigation is the way in which the water is sent to the plants via a channel or tunnel and sprayed at the end. But the major problem faced in this process is that, the amount of water wasted in this is too high. The problem solver that originated due to tiresome effort of many scientist is the Drip irrigation process.

What is Drip Irrigation Process?

Drip irrigation is a new and advanced process, that is not known by many. The way it differentiates from the other process is by, making the water directly available to the most needed part, rather than spraying it or converging a large amount through channels. This is achieved by making small openings in a pipe and focusing the water supply to the plants. The advantage associated with it is that it can save water to a large extent. The other thing about it is that it can even make a dry land into a fertile land. It is environmental friendly and to add to its advantages it can even challenge the nature itself. This was developed due to many hours hard work of many scientist. It is environmental friendly and accepted world wide for its capability. The limiting factor for this process, is the efficiency of the soaker hose.

What is Soaker hose and why it is Needed ?

All would have come across what a irrigation process is and what are the advantages associated with it. The item required to incorporate this method on the field is a soaker hose. The soaker hose is a item which resembles the structure of a normal garden hose. The difference it has over the garden hose is that, it has minute holes throughout its external surface. This forms the outlet for water to escape and irrigate the plants at particular places. The holes are so small that each hole is of the diameter of one by sixth to one by fourth of an inch. This is the working principle of a soaker hose. The unique advantage of the soaker hose over the other hoses is that it can fit with and pipes and fittings.

Maintenance of Drip Irrigation Systems

Agricultural practices have refined in last few years. People have broader idea about getting fruitful results out of their investment on the lands. One can ensure effective growth of plant only if the plant or whole plantation is administered with good irrigation facilities. These facilities can be helpful in providing useful gains for the farmers. By far technological advancements have gathered minds of farmers to render useful ideas into planned actions. These technological advancements can be adjudged with useful application in day-to-day life.

Importance of Irrigation System

Irrigation plays a vital role in the modern day agriculture. The root tips are in search of water when they travel down inside the ground. These can be ably supported by irrigation administered by the cultivators. The irrigation methods are of varied types that are available for cultivators. Starting from drip irrigation to centre pivot irrigation. Some of irrigation methods take care of all the necessary topological features that prevail in particular land. The contour of surface changes from place to place. It is quite evident that it can affect watering scheme adopted by the farmers. But this would be justified by the amount of water needed. Some crops would need large supply of water to be supplied to root tips.

Irrigation Facilities Promised by Drip Irrigation

Drip irrigation is one of the popular methods adopted by the farmers all over the world. This method could bestow useful irrigation to the crops that are planted. Drip irrigation is preferred in the gardens. The gardens that are small can be well sheltered under effective watering through drip irrigation scheme.

Also some watering scheme require some additional components such as tubing that can add to the cost of establishment of watering facility. This can be cashing for the farmers initially but with effective maintenance it can fetch huge profits on long run. Drip irrigation also ensures correct level of moisture that is to be maintained around the root tips. Any variation in this can cause adverse effects on the plant.

Drip Irrigations Systems – Components

Drip irrigation involves use of valuable wit to design the overall plan of the system. The system includes tubing and piping needs to be satisfied before going for commissioning of irrigation system. There are several components that form integral part of drip irrigation systems. Also one needs to ensure proper design of irrigation systems can be employed to facilitate moisture monitoring.

The shoot system of the plant needs supply of water and other nutrients through root. The watering of plants at optimum level ensures proper supply of nourishments for the development of crop into sizeable yielding. Special simulation software is available to map all the essential failure nodes that can lead to drastic changes in the irrigation channel. Hence it is advisable to use that software to clear out all the loopholes present in irrigation channel.

Effective Maintenance of Irrigation Systems

Maintenance of irrigation system largely depends on the design of irrigation system. The maintenance of the drip irrigation system involves series of actions that can ensure proper working of overall system. Subsurface drip watering schemes are widely used by the farmers pertaining to different parts of world to irrigate the lawns. These can be effectively monitored for pressure inside the channel. This pressure if not checked periodically can institute massive technical failures for the entire irrigation system. People can seek expert advice to have proper maintenance. The tubing can be checked for correctness periodically to eliminate water leakage. The inlet and outlet nozzle can be cleared off from dust and mud in order to have proper entry and exit of water inside the whole watering system.

Types of Drip Irrigation Valve

If you want to irrigate your farms with a drip irrigation system, having the right drip irrigation valves properly installed is a must.

There are various types of drip irrigation valve you can find in the market. Most drip irrigation systems need at least two different purposive-types: an emergency shut off one and a controlled one.

Emergency Shut Off Valves

Emergency shut off valves are drip irrigation valves to be installed in the area where you tap in for your drip irrigation system. It is the closest point to water source. Having these valves installed is crucial. Without them, you will have to shut off the water flow to your entire house should an irrigation breakdown occurs. Without them, you will also have to work on the mainline or irrigation valves.

The most commonly used for this purpose are gate valves. They are quite affordable though also tend to wear out quickly and then start leaking. Better recommended are ball valves, disk valves and butterfly valves. They are more reasonably pricey than gate valves since they are much more reliable and last several times longer.

Zone Control Valves

Zone control valves are drip irrigation valves that turn on and off water to the drip tubes (or hoses). Often these valves are automatically turned on and off by an irrigation controller/timer. For a small drip irrigation system you may need only one zone control valve. While for bigger drip irrigation systems several more may be required. There are two basic types of zone control valve: standard globe valve and anti-siphon valve.

Standard globe valves can be found in almost any size. They are often installed below the ground, kept inside a box or a vault. Since a standard globe valve does not incorporate a backflow preventer, you must provide one on your own.

Anti-siphon valves can be found in 20mm (3/4 inch) and 25mm (1 inch) sizes. They are most recommended for home gardeners since they incorporate a backflow preventer, thus saving a considerable amount of money. The valves must be installed above the ground, at least 150mm (6 inch) higher than the highest drip emitter. This requirement may be a problem for certain areas, but you can always install anti-siphon valves on top of trellis or at the top of the slope (if you have a garden with slopes).

Indexing Valves

Indexing valves are a single valve unit purposely installed to control several valve zones. Usually available in models with or without a built-in anti-siphon device, it requires a special controller to operate. It has a water inlet and several water outlets. When the control unit sends the first signal it opens the first water outlet. At the next signal it switches to the second water outlet. And then to the third, fourth and so on till it gets back to the first water outlet, at which point it shuts off. Indexing valves must be installed above the ground, at least 150mm (6 inch) higher than the highest drip emitter. It has never been too popular and is generally available in certain regions where a nearby manufacturer has promoted them.

Drip Irrigation Hoses Available in the Market

Drip irrigation hoses, also known as drip irrigation tubing, are hoses specifically designed to carry water and drip it through tiny holes attached to them; the tiny holes are called "emitters". They drip water slowly into the soil exactly at the plant root zone where it is needed. This way moisture levels are kept optimal, improving plants productivity and quality.

Drip irrigation hoses are made of polythene. They come in various types and diameters, accordingly to your needs. The length of a single drip irrigation hose should not over 200 feet from the point where water enters it.

You need to stake the hoses to keep them from moving. More importantly, never bury the hoses and their emitters even if they are designed to be. Otherwise, you will need to spend more time and energy to overcome clogging and rodent damage.

When choosing hoses for your drip irrigation system, you have to keep in mind these important factors: size, pressure rating, weight, length and chemical compatibility.

Some Drip Irrigation Hoses Available in the Market

Drip Irrigation Hose.580

It comes in various lengths: 25', 50', 100' and 500'.

Using.580 compression fittings, it operates at pressure rate 10 to 60 PSI and has a maximum flow rate 180 GPH. It is best if you operate it at pressure rate 25 PSI.

Drip Irrigation Hose.700

It comes in various lengths: 50', 100' and 500'.

Using.700 compression fittings, it operates at pressure rate 10 to 60 PSI and has a maximum flow rate at 240 GPH. It is best if you operate it at pressure rate 25 PSI.

Drip Irrigation Hose.820

It comes in various lengths: 100', 250' and 500'

Using.820 compression fittings, it operates at pressure rate 10 to 60 PSI and has a maximum flow rate at 380 GPH. It is best if you operate it at pressure rate 25 PSI.

1/43 Micro-Drip Irrigation Hose

It comes in various lengths: 25', 50', 100' and 500'. Used with 1/4" barbed fittings, it is used for extending drippers and micro-sprinklers from main line or as the primary line in a small drip irrigation system. It can be put above or below the ground. Use this hose to distribute water from main line to drippers, misters and low volume sprinklers.

1/43 Laser Drilled Drip Irrigation Hose

It comes in various lengths: 50' and 100'. Holes are laser drilled into the hose against the flow of water. Used with 1/43 barbed fittings, it is to be put within 12" or 6" space.

Advantages of a Subsurface Drip Irrigation System

Drip irrigation systems are mostly installed permanently or semi-permanently above the ground. When applied to irrigate trees and vine crops, they promote no problem. However, when applied to irrigate field and vegetable crops, certain problems occur, mostly because the drip irrigation systems have to be installed and retrieved annually. The annual handling is harmful to the life time span of the system, even under the best handling conditions. Hence the subsurface drip irrigation system emerges. The depth installment of subsurface drip irrigation system is determined by the soil type and the crops planted. An efficient installation has water moving at the depth of 4 to 30 inches beneath the surface, forming a continuous watered area along the plants row. Frequent irrigation cycles, up to several times every day, maximize capillary action and minimize water surfacing.

Four subsurface drip irrigation systems have been designed, installed and successfully produce better yields by the Water Management Research laboratory of the USDA in Fresno. The oldest one was installed in 1981 and used for three years to irrigate tomatoes. During those years, they also successfully produced twice as much broccoli, at the same farm. Subsurface drip irrigation system allows the precise application of water, nutrients and other agricultural chemicals directly to the plants root zones. It allows the optimization of growing environment, which leads to higher quality and quantity crop yields. With the same amount of water, subsurface drip irrigation system waters 46% larger volume of soil than any conventional (surface) drip irrigation system. It decreases the soil saturation, which not only leaves room for more air, but also improves water capillary movement and decreases water loss due to deep percolation. The top 15 – 20 cm of soil remains dry, therefore evaporation of water at the top of the soil may decrease and less salt will accumulate at the surface.

Permanent installation of subsurface drip irrigation system also provides considerable labour saving since irrigation can be applied while the whole equipment stays in the farm. Any soil surface crusts which usually lead to infiltration problems are no longer trouble makers; they are bypassed if this subsurface drip irrigation system is installed. Other farming applications required to support the conventional drip irrigation system is no longer required, too. Since subsurface drip irrigation system is installed below the ground, all components are not exposed to sun light, constant wetting, weather changes and other environmental changes. It is expected to last longer than the conventional drip irrigation system.

Drip Irrigation Greenhouses

Greenhouses type of plantation has gained massive support from the people all around. They need to have better growth under specified condition. Also the growth of plants can be termed healthy due to procurement of essential conditions that are for the uninterrupted growth of plants. The harmful radiations can be stopped from approaching the plants productive systems. Conducive conditions are reached at the greenhouses instituting better growth of plants.

Essentials for Drip Irrigation

Essentials for any type of irrigation system are water supply. The source of water can determine the type of irrigation employed. The drip irrigation involves effective management of the water through proper channel. The channeling of water can be established by rational drip irrigation design. Also the moisture level at the region beneath the soil surface to be regulated periodically with drip irrigation system

Why greenhouses require drip irrigation?

Drip irrigations are essential in area wherein there is scarcity in water.. Greenhouses are generally associated planned agriculture carried out at the expense of quite substantial investment. The effectiveness of the greenhouse is greatly determined by the quality of the plants grown so avoid any kind of pest or weed attack to the greenhouse plants, one needs to adopt for irrigation. It eliminates the opportunity for the growth of other types of plants like weeds taking nutrients from the ground surface.

Design your own Greenhouse Drip Irrigation

Drip irrigation can be customized based on the type of the field that is to be irrigated. The irrigation system involves the use of tubing, valves, pressure regulators and other plumbing accessories to get planned irrigation method. Pressure flow regulators to a maximum value of about 30 psi can regulate the operating pressure of tubing. The slope of field, soil features and water source can be carefully studied before validating a design. The water requirement for a particular variety of plant can determine particular type of accessory like emitter to be used.

Drip Irrigation Vegetable Gardens

Conventional irrigation system is confined with number of disadvantages to be associated with. The development of the facilities in agriculture has prompted people to adopt drip irrigation for carrying out all agricultural practices.

Drip Irrigation for Vegetable Gardens

Vegetable gardens are irrigated by drip irrigation in modern agricultural practice.. Gardens employ different kinds of plants to be grown under one shelter. This makes the work of gardener to be difficult in maintaining the garden. So it is advisable to use drip irrigation technique to have programmed watering of plants. This enables proper maintenance of moisture level in the ground. Since vegetable plants require different watering for obtaining better growth of plants.

Tips for Getting Efficient Drip Irrigation for Vegetable Gardens

It is essential to maintain adequate level of pressure in the tubing of drip irrigation. The micro irrigation technique can aid planned watering of younger plants. The watering of large and grown plants can be done using different types of emitters. Overall usage of water can be optimized with some practical testing of the drip irrigation system after installation. Vegetable plants like carrot can require only adequate amount of moisture to be maintained through out its growth cycle. So it is advisable for the designers to use multiple emitters and bubblers to carryout watering of those plants. Stream bubblers can be used for watering plants like tomato. The constant moisture can be preserved in the vegetable garden by drip irrigation technique.

Advantages of Drip Irrigation in Vegetable Gardens

Vegetable gardens with drip irrigation involve splitters used to channel the water for different plants. These can reduce the overall wastage of water yielding to effective water conservation. Drip irrigation provides random type watering enabling elimination of the weeds in vegetable garden. Vegetable gardens are prevented from soil erosion by using modular drip irrigation system.

Drip Irrigation Pressure Regulator

Drip irrigation is popular among all people residing all over the world. The utility of drip irrigation is characterized by the presence of certain implied conditions to be satisfied. These technical requirements determine the overall activity of the drip irrigation system.

Technical Requirements

The pressure is the main ingredient that is to be maintained by tubing with adequate level of supply of water. The tubing should withstand a pressure of about 2 bars. The tubing is to be maintained sufficiently deep in the ground. The ground can be dug to a depth that can be easily approached by the root tips of plants. Pressure regulation can be done to minimize or maximize the pressure that is maintained. Also flow valves can control the intensity of water. The ideal flow that is used is about 0.2l/s for standard dimension of tubing. The tubing that is generally employed in the drip irrigation is about 24mm diameter.

Pressure Regulator-Types

Pressure regulator is of different types. Varied range of pressure regulators can be used for the effective control of pressure in the tubing in different parts of the drip irrigation systems. They are numerous types and differ based on the design/specifications namely—

- Low flow pressure regulators. The design of low flow regulators is quite unique. Also the design can regulate pressure based on the utility. A flow rate of about 0.1 to 1GPM can be achieved with low flow pressure regulators.
- Medium flow pressure regulators. Medium flow regulators are utilized for the gardens that require the flow rate of range 2-20 GPM to be achieved. With this type of flow pressure regulators one can limit the pressure based on the inlet pressure applied.
- High flow pressure regulators. Flow rate can be limited to about 32 GPM with this type of high flow pressure regulators. The external pressure can vary from 10 psi till 50 psi.

Traditional Irrigation Fails Themoney Test

OUT OF the total 303.42 million ha of land that make up India, 136.18 million ha were categorised in 1987-88 as net sown area and of this, only 43.05 ha receives irrigation. The sources of irrigation, according to official records, are canals, tubewells, tanks, wells and other sources, a classification adapted by the British for the convenience of revenue categorisation. Nirmal Sengupta traces the evolution, decline, current status and future prospects of what has been categorised as tanks and 'other sources', which broadly, though not exhaustively, include wells, irrigation channels and lift irrigation schemes. These provide water for about 6.24 million ha. The book is fascinating both because of the subject with which it deals — user-friendly (participatory) irrigation systems — and the manner in which it has been treated. Narrating the schism in the usage of the terms traditional and modern, Sengupta desists from associating traditional with participation or user-friendliness though he makes no bones about the connection.

Old is Effective

He kicks off with the argument that "traditional" irrigation systems cannot be done away with because the terrain of our agricultural lands does not permit modern canals and tubewells to be effective in even one-fifth of the country. Therefore, future increase in irrigation capacity and even sustenance of current levels will depend on the expansion and maintenance of existing systems. The fact that "traditional" systems exist even today is a testimony to the sustainability, eco-viability and efficiency of these designs. Traditional techniques such as canal irrigation, which have been studied and developed by engineers of this era, have come to be labelled as "modern" systems. The process of selection has been based not on whether the system is traditional or modern, nor on the large-small divide; the dividing line is between people's participatory systems and centralised bureaucratic systems.

Volte-face

However, when it comes to the economic viability of these systems, Sengupta does a volte-face. He argues that financially, these "traditional" systems do not pass the test. The modern techniques, which were hitherto argued against so forcefully, suddenly become acceptable and the lacunae of them suffering from bureaucratic control can, it seems, be overcome by putting more effort into user participation while operating the system. At the end of it, the reader is left slightly confused. Is the book arguing for the current thrust in irrigation planning to be changed from mega-projects, which by necessity of design have to be centrally-controlled and unwieldy, to specific, user-need based projects, or is there a plea for smaller designs serving a smaller localised command and allowing people's participation? This is where Sengupta should have, but did not, draw a link between design, planning, control, management and social structure, and given his work a holistic perspective. Any water storage system or run-of-the-river facility has integrated links with the people it affects in more ways than one.

Any analysis of irrigation systems cannot be bereft of an understanding of the complex linkages on each of the above counts. Sengupta has, however, made no such attempt. Modern canal systems have linked with them issues such as displacement, resettlement and rehabilitation, changes in social and economic organisation in the local populace due to canal irrigation, agricultural practices and sustainability of resource use, which no sensitive study of irrigation can afford to bypass. While initially Sengupta does talk about the tremendous sustainability of user-friendly designs, he completely ignores this question while analysing the cost-benefit aspects of modern systems, understandably using a very narrow definition of costs, benefits and efficiency.

However, the problems notwithstanding, any serious researcher on Indian irrigation systems would necessarily have to add this book to his reading list.

7

Types of Irrigation

Various types of irrigation techniques differ in how the water obtained from the source is distributed within the field. In general, the goal is to supply the entire field uniformly with water, so that each plant has the amount of water it needs, neither too much nor too little.

Surface Irrigation

Surface irrigation is defined as the group of application techniques where water is applied and distributed over the soil surface by gravity. It is by far the most common form of irrigation throughout the world and has been practiced in many areas virtually unchanged for thousands of years.

Surface irrigation is often referred to as flood irrigation, implying that the water distribution is uncontrolled and therefore, inherently inefficient. In reality, some of the irrigation practices grouped under this name involve a significant degree of management (for example surge irrigation). Surface irrigation comes in three major types; level basin, furrow and border strip.

The Process

The process of surface irrigation can be described using four phases. As water is applied to the top end of the field it will flow or advance over the field length. The advance phase refers to that length of time as water is applied to the top end of the field and flows or advances over the field length. After the water reaches the end of the field it will either run-off or start to pond. The period of time between the end of the advance phase and the shut-off of the inflow is termed the wetting, pending or storage phase. As the inflow ceases the water will continue to runoff and infiltrate until the entire field is drained. The depletion phase is that short period of time after cut-off when the length of the field is still submerged. The recession phase describes the time period while the water front is retreating towards the downstream end of the field. The depth of water applied to any point in the field is a function of the opportunity time, the length of time for which water is present on the soil surface.

Basin Irrigation

Level basin irrigation has historically been used in small areas having level surfaces that are surrounded by earth banks. The water is applied rapidly to the entire basin and is allowed to infiltrate. Basins may be linked sequentially so that drainage from one basin is diverted into the next once the desired soil water deficit is satisfied. A "closed" type basin is one where no water is drained from the basin. Basin irrigation is favoured in soils with relatively low infiltration rates (Walker and Skogerboe 1987). Fields are typically set up to follow the natural contours of the land but the introduction of laser levelling and land grading has permitted the construction of large rectangular basins that are more appropriate for mechanised broadacre cropping. Basin irrigation is commonly used in the production of crops such as rice and wheat.

Furrow Irrigation

Furrow irrigation is conducted by creating small parallel channels along the field length in the direction of predominant slope. Water is applied to the top end of each furrow and flows down the field under the influence of gravity. Water may be supplied using gated pipe, siphon and head ditch or bankless systems. The speed of water movement is determined by many factors such as slope, surface roughness and furrow shape but most importantly by the inflow rate and soil infiltration rate. The spacing between adjacent furrows is governed by the crop species, common spacings typically range from 0.75 to 2 metres. The crop is planted on the ridge between furrows which may contain a single row of plants or several rows in the case of a bed type system. Furrows may range anywhere from less than 100 m to 2000 m long depending on the soil type, location and crop type. Shorter furrows are commonly associated with higher uniformity of application but result in increasing potential for runoff losses. Furrow irrigation is particularly suited to broad-acre row crops such as cotton, maize and sugar cane. It is also practiced in various horticultural industries such as citrus, stone fruit and tomatoes. The water can take a considerable period of time to reach the other end, meaning water has been infiltrating for a longer period of time at the top end of the field. This results in poor uniformity with high application at the top end with lower application at the bottom end. In most cases the performance of furrow irrigation can be improved through increasing the speed at which water moves along the field (the advance rate). This can be achieved through increasing flow rates or through the practice of surge irrigation. Increasing the advance rate not only improves the uniformity but also reduces the total volume of water required to complete the irrigation.

Surge Irrigation

Surge Irrigation is a variant of furrow irrigation where the water supply is pulsed on and off in planned time periods (e.g. on for ½ hour off for ½ hour). The wetting and drying cycles reduce infiltration rates resulting in faster advance rates and higher uniformities than continuous flow. The reduction in infiltration is a result of surface consolidation, filling of cracks and micro pores and the disintegration of soil particles during rapid wetting and consequent surface sealing during each drying phase. The effectiveness of surge irrigation is soil type dependent, for example many clay soils experience a rapid sealing behaviour under continuous flow therefore surge offers little benefit.

Bay/Border Strip Irrigation

Border strip or bay irrigation could be considered as a hybrid of level basin and furrow irrigation. The borders of the irrigated strip are longer and the strips are narrower than for basin irrigation and are orientated to align lengthwise with the slope of the field. The water is applied to the top end of the bay, which is usually constructed to facilitate free-flowing conditions at the downstream end. One common use of this technique includes the irrigation of pasture for dairy production.

Drainage after Harvest or in Rainy Season

Drainage of flooded banks or drainage of extremely wet soil during the rainy season may be done by ditches. Drainage by ditches may be done with crops that require the soil to be wet but not completely saturated (and sometimes, especially not at certain times of year). An example is blueberries. In the rainy season/winter, they require drier soil.

Issues Associated with Surface Irrigation

While surface irrigation can be practiced effectively using the right management under the right conditions, it is often associated with a number of issues undermining productivity and environmental sustainability

- Waterlogging-Can cause the plant to shut down delaying further growth until sufficient water drains from the rootzone. Waterlogging may be counteracted by drainage and watertable control.
- Deep drainage-Over irrigation may cause water to move below the root zone resulting in rising water tables. In regions with naturally occurring saline soil layers (for example salinity in south eastern Australia) or saline aqifers, these rising water tables may bring salt up into the root zone leading to problems of irrigation salinity.
- Salinization-Depending on water quality irrigation water may add significant volumes of salt to the soil profile. While this is a lesser issue for surface irrigation compared to other irrigation methods (due to the comparatively high leaching fraction), lack of subsurface drainage may restrict the leaching of salts from the soil. This can be remedied by drainage and soil salinity control.

Localized Irrigation

Localized irrigation is a system where water is distributed under low pressure through a piped network, in a pre-determined pattern, and applied as a small discharge to each plant or adjacent to it. Drip irrigation, spray or micro-sprinkler irrigation and bubbler irrigation belong to this category of irrigation methods.

Drip Irrigation

Drip irrigation, also known as trickle irrigation or micro irrigation, is an irrigation method which saves water and fertilizer by allowing water to drip slowly to the roots of plants, either onto the soil surface or directly onto the root zone, through a network of valves, pipes, tubing, and emitters.

History

Drip irrigation has been used since ancient times when buried clay pots were filled with water, which would gradually seep into the grass. Modern drip irrigation began its development in Afghanistan in 1866 when researchers began experimenting with irrigation using clay pipe to create combination irrigation and drainage systems. In 1913, E.B. House at Colorado State University succeeded in applying water to the root zone of plants without raising the water table. Perforated pipe was introduced in Germany in the 1920s and in 1934, O.E. Nobey experimented with irrigating through porous canvas hose at Michigan State University.

With the advent of modern plastics during and after World War II, major improvements in drip irrigation became possible. Plastic microtubing and various types of emitters began to be used in the greenhouses of Europe and the United States.

The modern technology of drip irrigation was invented in Israel by Simcha Blass and his son Yeshayahu. Instead of releasing water through tiny holes, blocked easily by tiny particles, water was released through larger and longer passageways by using velocity to slow water inside a plastic emitter. The first experimental system of this type was established in 1959 when Blass partnered with Kibbutz Hatzerim to create an irrigation company called Netafim. Together they developed and patented the first practical surface drip irrigation emitter. This method was very successful and subsequently spread to Australia, North America, and South America by the late 1960s.

In the United States, in the early 1960s, the first drip tape, called *Dew Hose*, was developed by Richard Chapin of Chapin Watermatics (first system established during 1964). In Pakistan it has been promoted by the Pakistan Atomic Energy Commission, the Agriculture Development Bank as well as successive governments. Beginning in 1989, Jain irrigation helped pioneer effective water-management through drip irrigation in India. Jain irrigation also introduced some drip irrigation marketing approaches to Indian agriculture such as 'Integrated System Approach', One-Stop-Shop for Farmers, 'Infrastructure Status to Drip Irrigation & Farm as Industry.' The latest developments in the field involve even further reduction in drip rates being delivered and less tendency to clog. One of the prestigious names in the field of drip irrigation in India is Kisan Irrigations Limited. Although among the top names in the field of Pipes & Fittings in India, they were a late entrant in this field. However, they have made important strides with their innovative & futuristic products like Hydrozig & Flat ranges in driplines as well as semi-portable, portable & flexible Mini and Micro Sprinkler Irrigation Systems for farmers in India. Their systems have been widely accepted by farmers due to the lowest clogging rate in operational conditions mainly due to the efficient Filtration Unit used which are indigenously manufactured at Kisan.

Modern drip irrigation has arguably become the world's most valued innovation in agriculture since the invention of the impact sprinkler in the 1930s, which offered the first practical alternative to surface irrigation. Drip irrigation may also use devices called micro-spray heads, which spray water in a small area, instead of dripping emitters. These are generally used on tree and vine crops with wider root zones. Subsurface drip irrigation (SDI) uses permanently or temporarily buried dripperline or drip tape located at or below the plant roots. It is becoming popular for row crop irrigation, especially in areas where water supplies are limited or recycled water is used for irrigation. Careful study of all the relevant factors like land topography, soil, water, crop and agro-climatic conditions are needed to determine the most suitable drip irrigation system and components to be used in a specific installation.

Components and Operation

Components (listed in order from water source):

- Pump or pressurized water source
- Water Filter(s)-Filtration Systems: Sand Separator like Hydro-Cyclone, Screen filters, Media Filters
- Fertigation Systems (Venturi injector) and Chemigation Equipment (optional)
- Backwash Controller (Backflow Preventer)
- Pressure Control Valve (Pressure Regulator)
- Main Line (larger diameter Pipe and Pipe Fittings)
- Hand-operated, electronic, or hydraulic Control Valves and Safety Valves
- Smaller diameter polytube (often referred to as "laterals")
- Poly fittings and Accessories (to make connections)
- Emitting Devices at plants (ex. Emitter or Drippers, micro spray heads, on-line drippers, trickle rings)
- Note that in Drip irrigation systems Pump and valves may be manually or automatically operated by a controller.

Most large drip irrigation systems employ some type of filter to prevent clogging of the small emitter flow path by small waterborne particles. New technologies are now being offered that minimize clogging. Some residential systems are installed without additional filters since potable water is already filtered at the water treatment plant. Virtually all drip irrigation equipment manufacturers recommend that filters be employed and generally will not honour warranties unless this is done. Last line filters just before the final delivery pipe are strongly recommended in addition to any other filtration system due to fine particle settlement and accidental insertion of particles in the intermediate lines.

Drip and subsurface drip irrigation is used almost exclusively when using recycled municipal waste water. Regulations typically do not permit spraying water through the air that has not been fully treated to potable water standards.

Because of the way the water is applied in a drip system, traditional surface applications of timed-release fertilizer are sometimes ineffective, so drip systems often mix liquid fertilizer with the irrigation water. This is called fertigation; fertigation and chemigation (application of pesticides and other chemicals to periodically clean out the system, such as chlorine or sulfuric acid) use chemical injectors such as diaphragm pumps, piston pumps, or venturi pumps. The chemicals may be added constantly whenever the system is irrigating or at intervals. Fertilizer savings of up to 95% are being reported from recent university field tests using drip fertigation and slow water delivery as compared to timed-release and irrigation by micro spray heads.

If properly designed, installed, and managed, drip irrigation may help achieve water conservation by reducing evaporation and deep drainage when compared to other types of irrigation such as flood or overhead sprinklers since water can be more precisely applied to the plant roots. In addition, drip can eliminate many diseases that are spread through water contact with the foliage. Finally, in regions where water supplies are severely limited, there may be no actual water savings, but rather simply an increase in production while using the same amount of water as before. In very arid regions or on sandy soils, the preferred method is to apply the irrigation water as slowly as possible.

Pulsed irrigation is sometimes used to decrease the amount of water delivered to the plant at any one time, thus reducing runoff or deep percolation. Pulsed systems are typically expensive and require extensive maintenance. Therefore, the latest efforts by emitter manufacturers are focused toward developing new technologies that deliver irrigation water at ultra-low flow rates, i.e. less than 1.0 liter per hour. Slow and even delivery further improves water use efficiency without incurring the expense and complexity of pulsed delivery equipment.

Drip Irrigation is Used by Farms, Commercial Greenhouses and Residential Gardeners

Drip irrigation is adopted extensively in areas of acute water scarcity and especially for crops such as coconuts, containerized landscape trees, grapes, bananas, ber, brinjal, citrus, strawberries, sugarcane, cotton, maize, and tomatoes.

Garden

Garden drip irrigation kits are increasingly popular for the homeowner and consist of a timer, hose and emitter. Hoses that are 4 mm in diameter are used to irrigate flower pots.

Advantage/Disadvantages

The advantages of drip irrigation are:

- Minimized fertilizer/nutrient loss due to localized application and reduced leaching.
- High water application efficiency.
- Levelling of the field not necessary.
- Ability to irrigate irregular shaped fields.
- Allows safe use of recycled water.
- Moisture within the root zone can be maintained at field capacity.
- Soil type plays less important role in frequency of irrigation.
- Minimized soil erosion.
- Highly uniform distribution of water i.e., controlled by output of each nozzle.
- Lower labour cost.
- Variation in supply can be regulated by regulating the valves and drippers.
- Fertigation can easily be included with minimal waste of fertilizers.
- Foliage remains dry thus reducing the risk of disease.
- Usually operated at lower pressure than other types of pressurised irrigation, reducing energy costs.

The disadvantages of drip irrigation are:

- Expense. Initial cost can be more than overhead systems.
- Waste. The sun can affect the tubes used for drip irrigation, shortening their usable life. Longevity is variable.
- Clogging. If the water is not properly filtered and the equipment not properly maintained, it can result in clogging.
- Drip irrigation might be unsatisfactory if herbicides or top dressed fertilizers need sprinkler irrigation for activation.
- Drip tape causes extra cleanup costs after harvest. You'll need to plan for drip tape winding, disposal, recycling or reuse.
- Waste of water, time & harvest, if not installed properly. These systems requires careful study of all the relevant factors like land topography, soil, water, crop and agro-climatic conditions, and suitability of drip irrigation system and its components.
- Germination Problems. In lighter soils subsurface drip may be unable to wet the soil surface for germination. Requires careful consideration of the installation depth.
- Salinity. Most drip systems are designed for high efficiency, meaning little or no leaching fraction. Without sufficient leaching, salts applied with the irrigation water may build up in the root zone, usually at the edge of the wetting pattern.

Dripperline

A dripperline is a type of drip irrigation tubing with emitters pre-installed at the factory.

Emitter

An emitter is also called a dripper and is used to transfer water from a pipe or tube to the area that is to be irrigated. Typical emitter flow rates are from 0.16 to 4.0 US gallons per hour (0.6 to 16 L/h). In many emitters, flow will vary with pressure, while some emitters are *pressure compensating*. These emitters employ silicone diaphragms or other means to allow them to maintain a near-constant flow over a range of pressures, for example from 10 to 50 psi (70 to 350 kPa).

Sprinkler Irrigation

In sprinkler or overhead irrigation, water is piped to one or more central locations within the field and distributed by overhead high-pressure sprinklers or guns. A system utilizing sprinklers, sprays, or guns mounted overhead on permanently installed risers is often referred to as a *solid-set* irrigation system. Higher pressure sprinklers that rotate are called *rotors* and are driven by a ball drive, gear drive, or impact mechanism.

Rotors can be designed to rotate in a full or partial circle. Guns are similar to rotors, except that they generally operate at very high pressures of 40 to 130 lbf/in^2 (275 to 900 kPa) and flows of 50 to 1200 US gal/min (3 to 76 L/s), usually with nozzle diameters in the range of 0.5 to 1.9 inches (10 to 50 mm). Guns are used not only for irrigation, but also for industrial applications such as dust suppression and logging.

Sprinklers may also be mounted on moving platforms connected to the water source by a hose. Automatically moving wheeled systems known as *travelling sprinklers* may irrigate areas such as small farms, sports fields, parks, pastures, and cemeteries unattended. Most of these utilize a length of polythene tubing wound on a steel drum. As the tubing is wound on the drum powered by the irrigation water or a small gas engine, the sprinkler is pulled across the field. When the sprinkler arrives back at the reel the system shuts off. This type of system is known to most people as a "waterreel" travelling irrigation sprinkler and they are used extensively for dust suppression, irrigation, and land application of waste water. Other travellers use a flat rubber hose that is dragged along behind while the sprinkler platform is pulled by a cable. These cable-type travellers are definitely old technology and their use is limited in today's modern irrigation projects.

Centre Pivot Irrigation

Centre-pivot irrigation (sometimes called central pivot irrigation), also called circle irrigation, is a method of crop irrigation in which equipment rotates around a pivot. A circular area centred on the pivot is irrigated, often creating a circular pattern in crops when viewed from above.

How it Works

Central pivot irrigation is a form of overhead (sprinkler) irrigation consisting of several segments of pipe (usually galvanized steel or aluminium) joined together and supported by trusses, mounted on wheeled towers with sprinklers positioned along its length. The machine moves in a circular pattern and is fed with water from the pivot point at the centre of the circle. The outside set of wheels sets the master pace for the rotation (typically once every three days). The inner sets of wheels are mounted at hubs between two segments and use angle sensors to detect when the bend at the joint exceeds a certain threshold, and thus, the wheels should be rotated to keep the segments aligned. Centre pivots are typically less than 500m in length (circle radius) with the most common size being the standard 1/4 mile machine (400 m). To achieve uniform application, centre pivots require a continuously variable emitter flow rate across the radius of the machine. Nozzle sizes are smallest at the inner spans to achieve low flow rates and increase with distance from the pivot point.

Most centre pivot systems now have drops hanging from a u-shaped pipe called a *gooseneck* attached at the top of the pipe with sprinkler heads that are positioned a few feet (at most) above the crop, thus limiting evaporative losses and wind drift. There are many different nozzle configurations available including static plate, moving plate and part circle. Pressure regulators are typically installed upstream of each nozzle to ensure each is operating at the correct design pressure. Drops can also be used with drag hoses or bubblers that deposit the water directly on the ground between crops. This type of system is known as LEPA (Low Energy Precision Application) and is often associated with the construction of small dams along the furrow length (termed furrow diking/dyking). Crops may be planted in straight rows or are sometimes planted in circles to conform to the travel of the centre pivot.

Originally, most centre pivots were water-powered. These were replaced by hydraulic systems and electric motor-driven systems. Most systems today are driven by an electric motor mounted at each tower.

For centre pivot to be used, the terrain needs to be reasonably flat; but one major advantage of centre pivots over alternative systems is the ability to function in undulating country. This advantage has resulted in increased irrigated acreage and water use in some areas. The system is in use, for example, in parts of the United States, Australia, New Zealand, Brazil and also in desert areas such as the Sahara and the Middle East.

Centre Pivot Manufacturers

There are four major centre pivot system manufacturers in the United States: Valmont Industries and their "Valley" products, Lindsay Corporation and their "Zimmatic" brand, Reinke Irrigation with their "Electrogator" machines, and T-L Irrigation who makes a hydrostatically powered system. Valley, Lindsay, and Reinke all manufacture systems powered by 480 volt electricity. T-L's variable-displacement hydraulic pump which is typically driven by a 15 horse power motor on standard quarter section irrigators. Water application typically consist of brass impacts, drip tubes, rotating nozzles, and stationary sprays. These sprinklers are manufactured by Nelson Irrigation and Senniger Irrigation. Varying applications, soils, and crops require different volumes of water and application rates. Pivots often have a large bore impact sprinkler (called "big guns") located on the vary end of the machine to aid in irrigating most number of acres possible. While these "end guns" may dramatically increase the irrigated area they suffer from poor uniformity and may have negative impacts on the entire pivot if not designed properly.

The largest maker of mechanized irrigation market is Valmont Industries. Valmont claims a market share between 55-65% of all new centre pivots sold in the United States. Reinke is a privately held company which limits the ability for market researches to determine the exact number of centre pivot sold. Reinke and Zimmatic compete to share between 30-40% of the irrigation market. Valley, Zimmatic, and Reinke manufacture modern irrigation equipment and consume about 95% and support networks of professional dealers. T-L Irrigation, also privately held, manufactures a hydraulically driven irrigator and typically sells through part-time and farmer dealers.

Linear/Lateral Move Irrigation Machines

The above mentioned equipment can also be configured to move in a straight line where it is termed a *linear move* or *lateral move* irrigation system. In this case the water is supplied by an irrigation channel running the length of the field and positioned either at one side or midway across the field width. The motor and pump equipment is mounted on a cart adjacent to the supply channel that travels with the machine. Farmers may opt for linear moves to conform to existing rectangular field designs such as those converting from furrow irrigation. Lateral moves are far less common, rely on more complex guidance systems and require additional management than compared to centre pivot systems. Lateral moves are common in Australia and typically range between 500-1000m in length.

Lateral Move (Side Roll, Wheel Line) Irrigation

A series of pipes, each with a wheel of about 1.5 m diameter permanently affixed to its midpoint and sprinklers along its length, are coupled together at one edge of a field. Water is supplied at one end using a large hose. After sufficient water has been applied, the hose is removed and the remaining assembly rotated either by hand or with a purpose-built mechanism, so that the sprinklers move 10 m across the field. The hose is reconnected. The process is repeated until the opposite edge of the field is reached. This system is less expensive to install than a centre pivot, but much more labour intensive to operate, and it is limited in the amount of water it can carry. Most systems utilize 4 or 5-inch (130 mm) diameter aluminum pipe. One feature of a lateral move system is that it consists of sections that can be easily disconnected. They are most often used for small or oddly-shaped fields, such as those found in hilly or mountainous regions, or in regions where labour is inexpensive.

Sub-irrigation

Subirrigation also sometimes called *seepage irrigation* has been used for many years in field crops in areas with high water tables. It is a method of artificially raising the water table to allow the soil to be moistened from below the plants' root zone. Often those systems are located on permanent grasslands in lowlands or river valleys and combined with drainage infrastructure. A system of pumping stations, canals, weirs and gates allows it to increase or decrease the water level in a network of ditches and thereby control the water table.

Sub-irrigation is also used in commercial greenhouse production, usually for potted plants. Water is delivered from below, absorbed upwards, and the excess collected for recycling. Typically, a solution of water and nutrients floods a container or flows through a trough for a short period of time, 10–20 minutes, and is then pumped back into a holding tank for reuse. Sub-irrigation in greenhouses requires fairly sophisticated, expensive equipment and management. Advantages are water and nutrient conservation, and labour-saving through lowered system maintenance and automation. It is similar in principle and action to subsurface drip irrigation.

Manual Irrigation Using Buckets or Watering Cans

These systems have low requirements for infrastructure and technical equipment but need high labour inputs. Irrigation using watering cans is to be found for example in peri-urban agriculture around large cities in some African countries.

Automatic, Non-electric Irrigation Using Buckets and Ropes

Besides the common manual watering by bucket, an automated, natural version of this also exist. Using plain polyester ropes combined with a prepared ground mixture can be used to water plants from a vessel filled with water.

The ground mixture would need to be made depending on the plant itself, yet would mostly consist of black potting soil, vermiculite and perlite. This system would (with certain crops) allow to save expenses as it does not consume any electricity and only little water (unlike sprinklers, water timers,...). However, it may only be used with certain crops (probably mostly larger crops that do not need a humid environment; perhaps e.g. paprikas).

Irrigation using Stones to Catch Water from Humid Air

In countries where at night, humid air sweeps the countryside, stones are used to catch water from the humid air by condensation. This is for example practiced in the vineyards at Lanzarote.

Dry Terraces for Irrigation and Water Distribution

In subtropical countries as Mali and Senegal, a special type of terracing (without flood irrigation or intent to flatten farming ground) is used. Here, a 'stairs' is made through the use of ground level differences which helps to decrease water evaporation and also distributes the water to all patches (sort of irrigation).

Sources of Irrigation Water

Sources of irrigation water can be groundwater extracted from springs or by using wells, surface water withdrawn from rivers, lakes or reservoirs or non-conventional sources like treated wastewater, desalinated water or drainage water. A special form of irrigation using surface water is spate irrigation, also called floodwater harvesting. In case of a flood (spate) water is diverted to normally dry river beds (wadi's) using a network of dams, gates and channels and spread over large areas. The moisture stored in the soil will be used thereafter to grow crops. Spate irrigation areas are in particular located in semi-arid or arid, mountainous regions. While floodwater harvesting belongs to the accepted irrigation methods, rainwater harvesting is usually not considered as a form of irrigation. Rainwater harvesting is the collection of runoff water from roofs or unused land and the concentration of this.

How an in-ground Irrigation System Works

Most commercial and residential irrigation systems are "in ground" systems, which means that everything is buried in the ground. With the pipes, sprinklers, emitters (drippers), and irrigation valves being hidden, it makes for a cleaner, more presentable landscape without garden hoses or other items having to be moved around manually. This does, however, create some drawbacks in the maintenance of a completely buried system.

Water Source and Piping

The beginning of a sprinkler system is the water source. This is usually a tap into an existing (city) water line or a pump that pulls water out of a well or a pond. The water travels through pipes from the water source through the valves to the sprinklers and emitters. The pipes from the water source up to the irrigation valves are called "mainlines", and the lines from the valves to the emitters or sprinklers are called "lateral lines". Most piping used in irrigation systems today are HDPE and MDPE or PVC or PEX plastic pressure pipes due to their ease of installation and resistance to corrosion. After the water source, the water usually travels through a check valve. This prevents water in the irrigation lines from being pulled back into and contaminating the clean water supply. Ideally a pressure control valve is also installed to regulate water pressure and help prevent excessive pressure from harming the system.

Controllers, Zones and Valves

Most irrigation systems are divided into zones. A zone is a single irrigation valve and one or a group of drippers or sprinklers that are connected by pipes or tubes. Irrigation systems are divided into zones because there is usually not enough pressure and available flow to run sprinklers for an entire yard or sports field at once. Each zone has a solenoid valve on it that is controlled via wire by an irrigation controller. The irrigation controller is either a mechanical (now the "dinosaur" type) or electrical device that signals a zone to turn on at a specific time and keeps it on for a specified amount of time. "Smart Controller" is a recent term used to describe a controller that is capable of adjusting the watering time by itself in response to current environmental conditions. The smart controller determines current conditions by means of historic weather data for the local area, a soil moisture sensors (water potential or water content), rain sensor, or in more sophisticated systems satellite feed weather station, or a combination of these.

Emitters & Sprinklers

When a zone comes on, the water flows through the lateral lines and ultimately ends up at the irrigation emitter (drip) or sprinkler heads. Many sprinklers have pipe thread inlets on the bottom of them which allows a fitting and the pipe to be attached to them. The sprinklers are usually installed with the top of the head flush with the ground surface. When the water is pressurized, the head will pop up out of the ground and water the desired area until the valve closes and shuts off that zone. Once there is no more water pressure in the lateral line, the sprinkler head will retract back into the ground. Emitters are generally laid on the soil surface or buried a few inches to reduce evaporation losses.

Problems in irrigation:

- Competition for surface water rights.
- Depletion of underground aquifers.
- Ground subsidence (e.g. New Orleans, Louisiana).

Underirrigation or irrigation giving only just enough water for the plant (e.g. in drip line irrigation) gives poor soil salinity control which leads to increased soil salinity with consequent build up of toxic salts on soil surface in areas with high evaporation. This requires either leaching to remove these salts and a method of drainage to carry the salts away. When using drip lines, the leaching is best done regularly at certain intervals (with only a slight excess of water), so that the salt is flushed back under the plant's roots.

- Over irrigation because of poor distribution uniformity or management wastes water, chemicals, and may lead to water pollution.
- Deep drainage (from over-irrigation) may result in rising water tables which in some instances will lead to problems of irrigation salinity.
- Irrigation with saline or high-sodium water may damage soil structure.

Reservoir

A reservoir is an artificial lake used to store water. Reservoirs are often created by building a reinforced dam, usually out of concrete, earth, rock, or a mixture across a river or stream. Once the dam is completed, the stream fills the reservoir. When a reservoir is predominantly man-made (rather than being an adaptation of a natural basin) it may be called a cistern. The term reservoir is also often used to describe underground reservoirs such as an oil or water well.

Types

Valley Dammed Reservoir

The more common dam across a valley relies on naturally formed features to form the watertight elements. Generally, engineers look for dam sites which are narrow with a broad area upstream; the valley sides can then act as natural walls and the broad area upstream makes a large reservoir for the height. The best place along the valley for building a dam has to be determined according to where the dam can best be tied into the valley walls and floor to form a watertight seal. If necessary, humans have to be re-housed or historic sites must be moved. For example, the temples of Abu Simbel were moved before the construction of the Aswan Dam (which created Lake Nasser from the Nile in Egypt). At the start of construction, the river must be diverted, often through a tunnel. Then the foundation is prepared. Once that is done, building of the dam can start. This may take anywhere from a few months to a few years, depending on its size and complexity. After the dam is complete, the diversion is removed or plugged, and the river fills the area upstream of the dam.

Bank-side Reservoir

Where water is taken from a river of variable quality or quantity, it is common to construct bank-side reservoirs to store water pumped or siphoned from the river. Such reservoirs are usually built partly by excavation and partly by the construction of a complete encircling bound or embankment. Both the floor of the reservoir and the bound must have an impermeable lining or core, often made of puddled clay. The water stored in such reservoirs may have a residence time of several months during which time normal biological processes are able to substantially reduce many contaminants and almost eliminate any turbidity. The use of bank-side reservoirs also allows a water abstraction to be closed down for extended period at times when the river is unacceptably polluted or when flow conditions are very low due to drought. The London water supply system is one example of the use of bank-side storage for all the water taken from the River Thames and River Lee with many large reservoirs visible along the approach to London Heathrow Airport.

Service Reservoir

Many service reservoirs are constructed as water towers, often as elevated structures on concrete pillars where the landscape is relatively flat. Other service reservoirs are entirely underground, especially in more hilly or mountainous country. In the United Kingdom Thames Water has many underground reservoirs beneath London built in the 1800s by the Victorians, most of which are lined with thick layers of brick. Honour Oak Reservoir, which was completed in 1909, is the largest of this type in Europe. The roof is supported using large brick pillars and arches and the outside surface is used as a golf course.

Operation

A raw water reservoir does not simply hold water until it is needed. It is the first part of the water treatment process. The time the water is held for before it is released is known as the *retention time.* This is a design feature that allows particles and silts to settle out, as well as time for natural biological treatment using algae, bacteria and zooplankton that naturally live within the water.

Water can be released from the reservoir, generally by gravity, to be cleaned for drinking water, generate electricity, or simply maintain the downstream flow. In the event that major rainfall occurs, water can be released via a spillway to avoid over-topping and compromising the integrity of the dam. Most modern reservoirs have a specially designed draw-off tower that can discharge water from the reservoir at different levels both to access water as the reservoir draws down but also to allow water of a specific quality to be discharged into the downstream river as compensation water.

Levels

The terminology for reservoirs varies from country to country. In the United States the normal maximum level of a reservoir lake is called *full pool,* while the minimum level it can function at is *dead pool.* The water below this point is also called the dead pool, while the water in between is called the *conservation pool.* Full pool may have different levels in summer and winter, or based on the local wet and dry seasons. Once a reservoir reaches dead pool, it is below the level at which the dam can release it downstream. At this point, the streambed beyond the dam goes nearly or completely dry, and electricity production stops. This is also often the point at which intakes for municipal water systems begin to suck air in, and must be extended into deeper water, where stagnant water quality is much poorer. This can be done either permanently with longer pipes, or temporarily with large hoses floated on small barges, such as until a severe drought or dam repairs are over.

Hydroelectricity

A hydroelectric power station consists of large turbines at the base of a dam. Water from the reservoir behind the dam is channelled through pipes and delivered to the turbines, which in turn, spin a generator to produce electricity.

Controlling Watercourses

Reservoirs can be used in a number of ways to control how water flows through downstream waterways.

Irrigation

Water in an irrigation reservoir is released into networks of canals mainly for use in farmlands or secondary water systems. Water in an irrigation reservoir is *generally* not used for drinking water, but in some cases is.

Flood Control

Commonly known as an "*attenuation*" or "*balancing*" reservoir, these are used to prevent flooding to lower lying lands, flood control reservoirs collect water at times of unseasonally high rainfall, then release it slowly over the course of the following weeks or months. Some of these reservoirs are constructed across the river line with the onward flow controlled by an orifice plate. When river flow exceeds the capacity of the orifice plate water builds behind the dam but as soon as the flow rate reduces the water behind the dam slowly releases until the reservoir is empty again. In some cases such reservoirs only function a few times in a decade and the land behind the reservoir may be developed as community or recreational land. A new generation of balancing dams are being developed to combat climate change. They are called "Flood Detention Reservoirs". Because these reservoirs will remain dry for long periods, there will be a question as to the stability of the clay core as it could dry out. A British company Instant Barrage Services has developed an interesting composite core fill as an alternative to clay, made from recycled materials that seems to work and has a much lower carbon footprint.

Compensation

If a standard reservoir is built on a river which is used as a source of power, a compensation reservoir may also be built to guarantee a sufficient flow of water downstream during the working hours of the water-powered industries.

Canals

Where a natural watercourse's water is not available to be diverted into a canal, a reservoir may be built to guarantee the water level in the canal; for example, where a canal climbs to cross a range of hills through locks.

Recreation

Reservoirs often provide for recreational uses. Most reservoirs are built for a civic purpose, but still allow fishing, boating, and other activities. At most reservoirs, special rules apply for the safety of the public.

Modelling Reservoir Management

There is a wide variety of software for modelling reservoirs, from the specialist Dam Safety Program Management Tools (DSPMT) to the relatively simple WAFLEX, to integrated models like the Water Evaluation And Planning system (WEAP) that place reservoir operations in the context of system-wide demands and supplies.

History

Five thousand years ago, the craters of extinct volcanoes in Arabia were used as reservoirs by farmers for their irrigation water. Dry climate and water scarcity in India led to early development of water management techniques, including the building of a reservoir at Girnar in 3000 BC. Artificial lakes dating to the 5th century BC have been found in ancient Greece. An artificial lake in present-day Madhya Pradesh province of India, constructed in the 11th century, covered 650 square metres (7,000 sq ft). In Sri Lanka large reservoirs have been created by ancient Sinhalese kings in order to save the water for irrigation. The famous Sri Lankan king Parakramabahu I of Sri Lanka stated " do not let a drop of water seep into the ocean without benefitting mankind ". He created the reservoir named Parakrama Samudra (sea of King Parakrama), which has astonished archeologists.

Measuring Irrigation Water

Effective irrigation water management begins with accurate water measurement. Water measurement is required to determine both total volumes of water and flow rates pumped.

Measurement of volumes will verify that the proper amount of water is applied at each irrigation and that amounts permitted by water management districts are not exceeded.

Measurement of flow rates will help to ensure that the irrigation system is operating properly. For example, low flow rates may indicate the need for pump repair or adjustment, partially closed or obstructed valves or pipelines, or clogged drip emitters. High flow rates may indicate broken pipelines, defective flush valves, too many zones operating simultaneously, or eroded sprinkler nozzles.

Units of Measurement

Volume (V)

Volume is the amount of water measured. For irrigation purposes, volume units commonly used are the gallon, acre-inch, and acre-foot. An acre-inch is the volume of water that would be required to cover an area of 1 acre to a depth of 1 inch. The relationships among these units are:

1acre-inch (ac-in)= 27,154 gallons (gal)

1acre-foot (ac-ft)=12 acre-inches (ac-in).

Depth (D)

Depth units are sometimes used to refer to the amount of water required for irrigation. Depth units (inches) are used because soil water-holding capacity is typically measured in inches (of water) per foot (of soil depth), and irrigations are scheduled after a fraction of the soil water in the plant root zone has been depleted.

For example, assume that a typical Florida fine sand soil holds 1.0 inch of water per foot of soil at field capacity, the irrigated plant root zone depth is 2 ft, and an irrigation will be scheduled when 50% of the soil water in the root zone has been depleted. Then, the total soil water storage is calculated as 2.0 inches (D = 1.0 in/ft x 2 ft soil depth = 2.0 inches), and the amount to be applied per irrigation is 1.0 inch (D = 50% x 2 inches = 1.0 inch).

Note that the amount of water to be pumped will need to be greater than the 1.0 inch to be stored in the plant root zone because some water will be lost during application. That is, application efficiencies are always less than 100 percent because of water losses due to such factors as evaporation, wind drift, and nonuniform water application. Depth units are also convenient for comparison with rainfall depths. For example, a 1-inch rainfall would supply the same volume of water as the 1-inch irrigation in the previous example. Either the 1-inch rain or irrigation would be adequate to restore the soil water content to field capacity.

Finally, plant water use rates are typically expressed in (fractions of) inches per day. For example, if the plant ET rate in the previous example is 0.25 in/day, then the allowable soil water depletion of 1.0 inch will occur in four days (1.0 inch/0.25 inches/day = 4 days), and an irrigation would be required beginning on day 5 to avoid depleting more than 50% of the available soil water in the plant root zone.

The calculation of 1.0 inch of water to be applied means that a depth of 1.0 inch is to be applied over the entire area to be irrigated. If the irrigated area is 1.0 acre, then the volume to be applied is 1.0 ac-in (V = 1.0 inch depth x 1.0 acre = 1.0 ac-in) or 27,154 gal. Likewise, if the irrigated area is 20 acres, the volume of water required is 20 ac-in or 543,080 gal (V = 20 ac-in x 27, 154 gal/ac-in = 543,080 gal). Thus, depth units are used interchangeably with volume units because it is convenient to use depth units when referring to soil water depletion, rainfall, and plant ET rates, but volume units when referring to amounts of water permitted or pumped.

Flow Rate (Q)

Flow rate is the volume of water flowing past a given point per unit of time. The units of flow rate commonly used for irrigation purposes are gallons per minute (gal/min or gpm) and acre-inches per hour (ac-in/hr). The water management districts often use million gallons per day (mgd) for permitting of water extraction rates from a water source. The relationships between these units are:

1 gal/min = 0.00221 ac-in/hr = 0.00144 mgd

1 ac-in/hr = 453 gal/min = 0.652 mgd

1 mgd = 694 gal/min = 1.53 ac-in/hr.

Since flow rate is volume per unit time, the volumes of water applied during irrigation can be calculated if the flow rate and irrigation duration (time) are known. For example, if an irrigation pump discharges 453 gpm and irrigates for 6 hours, the volume of water applied is 163,080 gal (V= 453 gal/min x 60 min/hr x 6 hr = 163,080 gal). This is equivalent to 6.0 ac-in (V = 163,080 gal/27,154 gal/ac-in = 6.0 ac-in).

Flow rates can be converted to water application rates over the irrigated area if the size of the irrigated area is known. Water application rates from sprinkler irrigation systems are typically given in in/hr, consistent with rainfall rates. If, in the previous example, a sprinkler irrigation system applies water at 453 gpm, and the irrigated area is 4 acres, then the application rate is 0.25 in/hr (453 gpm = 1.0 ac-in/hr, and 1.0 ac-in/hr/4 acres = 0.25 in/hr).

Finally, the depth of water applied can be calculated by multiplying the application rate by the duration of irrigation. Continuing the previous example, for an application rate of 0.25 in/hr and a 6 hr irrigation duration, the gross depth of water applied would be 1.5 inches (D = 0.25 in/hr x 6 hr=1.5 inches). Note that the actual depth of water stored in the plant root zone and available for plant use would be less than 1.5 inches because of water losses during application. A typical application efficiency for sprinkler irrigation systems is 75% for Florida conditions. Thus, only 1.13 inches (1.5 inches x 75% = 1.13 inches) of the 1.5 inches pumped would be expected to be stored in the root zone and available for plant use.

Velocity (v)

Velocity is the average speed at which water moves in the direction of flow. The velocity unit commonly used is feet per second (ft/sec or fps). To aid in understanding the size of this unit, it is compared with a more familiar unit:

1 ft/sec = 0.68 miles per hour (mph)

To avoid excessive pressure losses due to friction and excessive potentially damaging surge pressures, most irrigation systems are designed to avoid velocities that exceed 5 ft/sec:

5 ft/sec = 3.4 mph.

Irrigation water velocities and flow rates are sometimes mistakenly used interchangeably. Velocity and flow rate are two different (but related) concepts. Their relationship is given by the equation of continuity, a fundamental physical law:

$$Q = a\ v$$

where

Q= flow rate,

a = cross-sectional area of flow, and

v = velocity.

The equation of continuity states that the flow rate can be calculated from the multiple of the velocity times the cross-sectional area of flow.

In pipes flowing full, the cross-sectional area of flow is the cross-sectional area of the pipe. In open channel flow, it is the cross sectional area of the channel or stream. Thus, the equation of continuity simply states that in addition to the velocity of flow, the size of the stream or pipe also affects the flow rate.

As an example of the application of the equation of continuity, if the velocity is the same in a 1-inch and a 2-inch diameter pipe, the flow rate from the 2 inch pipe would be four times as large as the flow rate from the 1-inch diameter pipe. Note that the cross-sectional area is proportional to the diameter squared and 1 inch squared = 1 inch x 1 inch = 1 square inch, while 2 inches squared = 2 inches x 2 inches = 4 square inches. From this example, doubling the pipe diameter increases the carrying capacity of a pipe by a factor of 4.

As another example, if the previously-described 1-inch and 2-inch diameter pipes must both convey the same flow rate, then the velocity in the 1-inch pipe will be 4 times greater (inversely proportional to the ratio of the cross-sectional areas) than the velocity in the 2-inch pipe.

Summary

The amount of irrigation water pumped, applied, or used can be measured in volume or depth units. These units can be used interchangeably when it is convenient to do so. That is, the volume can be calculated from the depth applied if the area is known over which the depth was applied.

The pumping rate, application rate, or plant water use rate is measured in units of volume (or depth) per unit of time. The volume can be calculated from the flow rate times the duration (time) of flow. The depth of application or use can be calculated from the application or use rate times the duration (time) of application or use.

Velocity refers to the speed at which water flows (distance per unit time) as opposed to flow rate (volume per unit time). Thus, the terms flow rate and velocity cannot be used interchangeably. Both the velocity and cross-sectional area of flow must be known in order to calculate flow rates.

8

Modern Irrigation Techniques

In order to feed very large numbers of people, modern farming has become very efficiency-conscious. Every possible way of improving crop yields is pursued. Very large, complex, expensive equipment is now virtually required, in order for a farm to be profitable. This also means that extremely large farms must be used, to amortize the very high cost of such equipment. It also means that massive use of pesticides and herbicides is necessary to process such huge areas regarding insects and weeds. More to the point here, phenomenal amounts of water is needed to permit all those plants to grow at optimal rates.

Natural soil moisture and natural rain used to be the sources that were relied on. Modern farming cannot rely on the natural variations of such things. Irrigation is central to all modern farming.

Considering how cost-conscious and efficiency-conscious modern farming has become, it amazes me at how poorly irrigation is used! Farmers tend to operate their irrigation pumps while they are awake, during the daytime. That seems to make sense, for the possibility of any mechanical malfunction. But for several reasons, it is very poor use of that resource.

- First, during hot, sunny summer days, a significant amount of the sprayed water evaporates in the warm air between the time it leaves the spray heads and gets to the plants.
- Second, once water lands on the leaves of the plants, the bright sun and hot temperatures causes even much more evaporation of that water, before it can be of any use to the plants.
- Third, due to the surfaces of the leaves, many of the drops of irrigation water rest on the leaves as relatively spherical drops

> and not as a uniform wetting of the surfaces. This is especially true when a plant first begins to be sprayed. Unfortunately, one effect of this droplet shape is to optically focus the sunlight, very much as a glass lens would do. In the same way that a glass magnifying lens concentrates sunlight so much that it can quickly burn a piece of paper, these droplets can cause very small areas of the leaf surfaces to receive too much sunlight! Cell damage can result, even though later droplets of water are likely to cool it down to ameliorate this sunburn effect.

Regarding this third issue, many homeowners already know that it is a bad idea to water their lawns in the middle of a very hot, sunny, summer day, because the lawn can become "burned". Why don't the technological modern farmers know that?

An obvious solution to this exists! If irrigation was begun in the early evening, and potentially continued to a little after sunrise, fully twelve hours of watering is practical each day. FAR less water would be required, due to the much lower evaporation that occurs during the cooler, sunless nights. The plant leaves would similarly not have to endure localized focused areas of sunlight and the possible damage that can result from that.

The benefits from this simple change can be significant. First, from a practical view, irrigation pumps would need to run fewer hours to provide equivalent usable moisture for the plants, which would lessen equipment maintenance costs and the cost of the electricity to run those very large irrigation pumps. Second, if less water is needed to fulfil the needs of the plants, less water would be removed from the underground aquifers.

On this last matter, there seems to have long been an incorrect assumption that the amount of available underground water is unlimited. It is not. In the American Midwest, there are many thousands of cities and towns that drilled wells into a large aquifer, whose water seems to have fallen in the Colorado region. This water flows horizontally rather slowly, and it apparently takes hundreds or thousands of years to seep those many miles across the country. Since there are now thousands of agricultural irrigation wells that also draw water from this same aquifer, a problem is developing.

In recent years, it has been becoming obvious that aquifer is NOT of infinite capacity. Due to all these municipal and irrigation wells drawing so much water out, the level of the water in the aquifer has been dropping, and rather rapidly! Many wells that have operated properly for many decades have become "dry". The solution is now to have those wells drilled deeper, and many cities and towns and homes have had to do just that in recent years. But the pattern is alarming. If wasteful and extravagant use of fresh water continues, there will some day be a time when, no matter HOW deep those wells will be drilled, there will be no available water in that aquifer. Considering that water to replenish it may take hundreds or thousands of years to seep through the many miles of rocks to get there, a VERY serious problems seems likely to exist!

Now, no one knows how much that aquifer initially had, or to what percent it has already been depleted. There is not even very good data on the collective total of water that is being removed from it! Each farm considers the water under it to be "personal property" which can be used in any ways and in any amounts desired! Various States attempt to keep track of Municipal water use, but the data on irrigation consumption is very sketchy.

In any case, we can hope that we have only depleted 10% of that aquifer so far, which would imply that the millions of people who live in the Midwest can expect to have supplies of fresh water for at least a couple hundred more years, before the aquifer will be so depleted that Municipal water will no longer be available. But what if the massive consumption of water in the past 30 years has 50% depleted it? That would mean that crisis could occur in only another 30 years! Many people living now would be faced with a total lack of available fresh water. Catching rainwater might enable survival, but rainwater contains many contaminants that are picked up from the air as the droplets fall from the clouds, so it is not really a "safe" source of drinking water.

If, when, this crisis happens, there will really be NO solution! All of those farms that REQUIRE massive amounts of irrigation will no longer be able to operate. The farmers will likely just haul their equipment to Brazil or some other country where available farming land will still have available irrigation water. But the homeowners and the business owners and the towns, what will they be able to do? How could they ever live and work, if fresh water would not be available for hundreds of years?

I realize that, as long as politicians do not see a looming crisis on this issue, that nothing will be done. And even if new laws would be passed, how could anyone monitor the operation of wells on the PRIVATE LAND of the many farms? Considering that farmers know that they MUST use irrigation water in order to be profitable, is there even the remotest chance that they would just stop?

Can Irrigation be Sustainable

Globally about 10 Mha of agricultural land are lost annually due to salinisation of which about 1.5 Mha is in irrigated areas. While some climate and management aspects are common to semi-arid regions the detailed mechanisms and options to secure ecological sustainability and economic viability may vary considerably from case to case. This paper applies a whole of system water balance to compare irrigation in three semi-arid regions suffering from similar sustainability issues: Rechna Doab (RD)-Pakistan, the Liuyuankou Irrigation System (LIS) – China and Murrumbidgee Irrigation Area (MIA)-Australia. Soil salinity, lack of adequate water resources and groundwater management are major issues in these areas. The MIA and LIS irrigation systems also suffer from soil salinity and low water use efficiency issues. These similarities occur in spite of very different climatic and underlying hydrogeological conditions. The key data used to compare these different regions are climate and soils, available water resources and their use, as well as components of the water balance. In addition, the history of water resource development in these areas is examined to understand how salinity problems emerge in semi-arid regions and the consequences for production. Based on the efficiency parameters and the definitions of sustainability, approaches are explored to solve common environmental problems while maintaining economic viability and environmental sustainability for irrigation systems.

Media Summary

It is possible to maintain the productive function of any irrigation area by providing adequate drainage and salt export facilities that may, however require high energy and capital investments. The most cost-effective option for sustainable irrigation is to increase water use efficiency and minimize negative impacts on the environment. There is a need to radically rethink sustainability of food production, rational pricing and sharing of water and commodities to justify the investment required to maintain and enhance ecosystem function within irrigated catchments.

Just 20 % of the world's croplands are irrigated but they produce 40 % of the global harvest which means that irrigation more than doubles land productivity (FAO, 2003). In developing countries irrigation improves economic returns and can boost production by up to 400%. On the other hand, irrigation can have unwanted environmental consequences. About one-third of the world's irrigated lands have reduced productivity as a consequence of poorly managed irrigation that has caused water logging and salinity (FAO, 1998).

Irrigation has been important for agricultural production in Mesopotamia (parts of present day Iraq and Iran) for 6000 years. The region has low rainfall and is supplied with surface water by two major rivers, the Tigris and the Euphrates. The plains of Mesopotamia have always had problems with poor drainage of soils, drought, catastrophic flooding, silting, and soil salinity. Although Mesopotamia is very flat, the bed of the Euphrates is higher than that of the Tigris; in fact, floods of the Euphrates sometimes found their way across country into the Tigris. Engineers took advantage of this gradient as soon as irrigation schemes became large enough, by using the Euphrates water as the supply and the Tigris channel as a drain.

The main engineering problems for earlier civilisations were water storage, flood control and maintenance of canals. The salinity problem was more subtle, not fully appreciated, and could not be overcome by the engineering available at the time. It was difficult to drain water from fields, and there was always a tendency for salt to accumulate in the soil.

The problems of irrigated agriculture in Mesopotamia can be summarised as:

- *Silting of canals:* silt built up quickly in the canal beds, threatening to block them
- *Soil salinity*: recorded evidence around 2000 BC, 1100 BC, and after 1200 AD
- *Water politics* arising from tension between upstream and downstream users. In Sumeria, the city of Lagash was far downstream in the Euphrates canal system. The governor of Lagash apparently decided that he would cut a canal to tap Tigris water rather than rely on water from the Euphrates, but the addition of poor-quality water from the Tigris led to rapid salinization of the soil.
- *Over exploitation of resources*: After the wave of Moslem expansion overtook Mesopotamia, the Abassid Caliphate was based in Baghdad from 762 AD until its demise in 1258. Existing irrigation schemes were renovated and greatly extended in very large projects. Abassid engineers drew water

from the Euphrates at five separate points, and led it in parallel canals across the plains, watering a huge area south of Baghdad. This system provided the basis for the enormously rich culture of Baghdad, which is still remembered in legend (Scheherezade, the Caliph of Baghdad, and the Arabian Nights) as well as history. But the scheme required a high level of physical maintenance, and there was increasing salinisation in the south.

- *Institutional failure*: As the central government began to fail in the 12th century (mostly from overspending), the canals became silt-choked, the irrigation system deteriorated, and the lands became more salinised. The deathblow to the system was aided by nature: massive floods about 1200 AD shifted the courses of both the Tigris and the Euphrates, cutting off most of the water supply to the Nahrwan Canal and wrecking the whole system. The Abbasids were too weak (or bankrupt) by then to institute repairs, and the agricultural system collapsed. By the time the Mongols under Hulagu devastated Iraq and Baghdad in 1258 AD, they conquered a society that occupied wasteland. Iraq has remained a desert for more than 600 years.

Key Challenges to Irrigated Agriculture

About 50% of the total developed fresh water resources of Asia are devoted to growing rice (Barker et al., 2001). In Asia, current estimates show that by 2025, 17 million hectares (Mha) of the irrigated rice area may experience "physical water scarcity" and 22 Mha "economic water scarcity". It is projected that global rice consumption in 2020 will increase by 35% from the levels of 1995, whereas water availability for agriculture over this period is expected to fall from 72 to 62% globally and from 87 to 73% in developing countries. Increasing water scarcity threatens the sustainability of irrigated agriculture and hence food security and the livelihoods of rice producers and consumers.

Increasing competition from domestic and industrial uses has further compounded the problem of water scarcity. The demand for freshwater for industrial and domestic urban needs is growing rapidly throughout Asia. Less water will be available for agriculture and for rice, the crop that consumes the largest amount of freshwater. In some areas, water scarcity is already a major problem and a serious limit to agricultural development. Farmers are under pressure to grow more "crop per drop". Therefore there is an urgent need to find ways to "grow more rice with less water", to achieve this, efficient and appropriate irrigation technologies are needed.

Some of the challenges faced by present day irrigated agriculture are similar to ancient Mesopotamia and are summarised below:

- *More efficient use of inputs* (water, fertilizer, pesticides and labour) aimed at reducing negative impacts on the environment and to reduce production costs.
- *Soil salinity*: Matching landscape capability with irrigation systems, there is a need to manage salinity hazard by matching landscape capability when making decisions on new locations for irrigation development or on-farm field suitability.
- *Minimising environmental impacts*: Management of extraction impacts by quantifying both positive and negative externalities of different irrigation areas and sectors by evaluation, auditing and benchmarking in the irrigation industry. Management of negative environmental impacts, such as methane and nitrous oxide emission, salinity, water pollution (abuse of pesticides), algal blooms etc., especially in intensive crop production systems.
- *Balancing consumptive and environmental demands*: Balancing irrigation and environmental flow demands through real savings due to improved distribution and on-farm water use efficiency and alternative cropping options.
- *Maintaining and enhancing quality* of drainage water and minimising impacts on rivers and ecosystems.
- *Institutional robustness*: Avoid a repeat of Mesopotamia in terms of institutional and ecosystem failures through better definition of property rights.
- *The Challenge:* to improve water use efficiency in rice-based irrigation systems of Australia and Asia

Three Regional Irrigation Perspectives in Rice-Based Systems

In 2000, the world's rice production was about 600 million tones Mt), 91 of which was produced in Asia. China, India, Indonesia, Bangladesh, Vietnam, Thailand, Myanmar, Philippines, Japan, and Brazil are the top 10 rice-producing countries (FAO, 2000).

Australia

In Australia, the rice crop is mainly grown on the Riverine Plain in New South Wales by utilizing surface water supply from the Murray and Murrumbidgee rivers and pumping from the Murray Aquifer System. Water is mainly supplied to the crop through a channel network serviced by irrigation companies or is pumped by the farmers directly from the rivers and creeks. According to the Australian Bureau of Statistics, gross water supplies in Australia to the rural sector (mainly used for irrigation) increased from 12.7 BCM (billion cubic metres) in 1983–84 to 15.8 BCM in 1995–96. Irrigated areas increased by 3.7% per year from 1.627 Mha in 1983–84 to 2.332 Mha in 1993–94.

Water use is currently 72% of the total water used for Irrigation in Australia. Surface water diversions in the Murray Darling systems are limited to a total of 11.6 BCM out of which around 1.6 BCM are used in rice production. All operations are mechanized. From 1.2 to 1.6 Mt of paddy rice are produced per year on 2300 farms. In 2001, 1.7 Mt of rice were harvested from an area of 184,000 ha (The Rice Marketing Board for the State of New South Wales 2001). The industry has a farm-gate value of approximately $350 million (AUD) and total value (export earnings, value-added) of over $800 million (Sunrice, 2002).

Including flow-on effects, it is estimated that the industry generates over $4 billion annually benefiting regional communities and the Australian economy. Normally the crop is grown in pounded water from sowing or from the 3-leaf stage. Rice is often grown in rotation with leguminous pastures and dryland crops, which improve soil fertility and limit the need for pesticides. The number of farms growing rice is restricted because there is limited water available for irrigation. Due to the increasing demand and competing water uses, state and federal governments have established policies to control the allocation of water to all users and to meet necessary environmental requirements.

The ability of the soil to pond water without excessive accessions to the groundwater or environmental effects to other lands are factors considered in approving areas of a farm suitable for rice production. The drainage water from paddy fields is recycled to maximize the utility of irrigation water and minimize off-farm effects of the irrigation system. Soil salinisation due to rising water tables in the rice-growing areas is a major concern for the rice based farming systems in the region. Other concerns include the volume and chemical composition of drainage waters, breeding of mosquitoes and bird control. To reduce water requirements of the rice crop, researchers are trying to shorten the growth duration and increase cold tolerance of the crop to minimize pounding duration.

Pakistan

The cultivated area in Pakistan is about 20 Mha of which more than 16 Mha are irrigated. About 11 Mha of the irrigated area (i.e. 73 % of the total) is situated in Punjab Province which is in the rice-wheat agro-ecological zone of Pakistan.

The Indus Basin Irrigation System in Pakistan is the largest integrated irrigation system in the world. This irrigation system diverts approximately 123 BCM of annual river flow and spreads it over 13.5 Mha of cultivable land, of which nearly 9 Mha can be irrigated throughout the year. This controlled distribution is accomplished by means of 17 barrages and canal diversion works, 42 major canals, 6,000 km of minor canals, 600 km of link canals, and 78,000 watercourses. The total capacity is nearly 7,000 m^3/sec.

This flow is supplemented with over 150,000 tube wells, which pump 24.5 BCM/year from groundwater. Rice cultivation in Pakistan is concentrated in the central Punjab and north-western districts of Sindh, where both surface and groundwater irrigation systems are well developed. Basmati rice is the principal cash crop in the Kharif (summer) season and wheat in the Rabi (winter) season. Rice occupies about 25 % of the cultivated area in the summer monsoon season and 10 % of the total cropped area. Wheat, being the staple food, occupies 75 % of the cultivated areas in the winter season and about 38% of the total cropped area. Pakistan is among the four major rice-exporting countries, but produces only 5 million tons compared to Bangladesh (34); Myanmar (21); India (132); Japan (11); Philippines (13); Thailand (26); Vietnam (32); China (182) and the world (581 Mt).

In Pakistan, rice transplanting coincides with the onset of monsoon rains, which meet the major portion of the rice water requirement. Pakistan has a huge potential to increase rice growing on a large scale due to its relatively level terrain, heavy soils with good water holding capacity, sunny days, appropriate climatic conditions and abundant supply of farm labour. Unfortunately, inadequate supply of irrigation water at critical times of growth, lack of drainage, saline and sodic soils, low quality seeds, antiquated farm implements, imbalances in farm inputs, unsatisfactory agriculture and irrigation practices are major constraints limiting rice area growth and generally crop production in Pakistan. Since the introduction of canal irrigation, waterlogging and soil salinity have become the major problems impeding agricultural growth and development.

China

Asia produces 90 % of the world's rice in a climatic zone which can have an annual precipitation of more than 1,500 mm. China is the major rice producing country having about 30 Mha under rice paddy cultivation with a total rice yield of 190 Mt, which is 32% of world rice production. China has a long history of irrigation. Two thousand years ago, the world-famous Dujiangyan irrigation district in Sichuan Province was built and remains in use after many periods of rehabilitation and modification.

With a current irrigated area of 670,000 ha it is a highly developed economic centre of great importance to China's food production. In 1949, the irrigated area of the whole country was 16 Mha accounting for only 16% of the country's farmland, and the per-capita food consumption was 209 kg. Development in the last half century has resulted in the irrigated area growing to 53 Mha, accounting for 40% of the farmland, and the per-capita food consumption was 400 kg by the end of 1998. The total water use for irrigation gradually increased from approximately 100 BCM in 1949 to 358 BCM in 1980, after which it stabilised. Irrigation water use has been stable since 1980. Industrial and municipal water use has increased rapidly and reduced the proportion used in irrigation to 92, 80, and 65% in 1949, 1980 and 1997, respectively. The efficiency of irrigation water use and the production efficiency have progressively risen over the past several decades, especially since 1980. The average water use for irrigated agricultural was 875 mm in 1980 decreasing to 780 mm in 1997. The average specific food yield during the same period increased from 0.6 kg/m^3 to 1 kg/m^3.

Rice contributes over 39% of the total food grain production in China from 31 Mha (28% of 113 Mha which is the total agriculture area). In 1999, the total rice yield reached 200 Mt, which accounts for 39% of the total national grain production of the country (Editorial Committee of the Year Book of Chinese Agriculture, 2001). Before the 1970s, the traditional irrigation regime for rice was "continuous deep flooding irrigation". Under this regime, a low yield of rice was obtained with a large amount of water. As the industrial, urban and rural domestic water consumptions have increased continuously, there are less water resources available for irrigation year by year. These pressures on the water supply are similar to those experienced around the world and have raised the importance of water worldwide. Recognising the need to improve both water and land productivity China has introduced a Water Efficient Regimes programme (WEI) for their rice industry. Since the introduction of the programme in the 1980s many regions have adapted one of the three WEIs:

- S.W.D. Combining shallow water depth with wetting and drying
- A.W.D. Alternate wetting and drying
- S.D.C. Semi dry cultivation.

By 1997, 5.7 Mha had been converted to one or other of these systems. A reduced environmental impact has been claimed for each method.

Climatic, Water Balance and Hydrological Comparison of Three Subsystems

The key water balance components are compared for three irrigation areas, viz., Murrumbidgee Irrigation Area (MIA), Rechna Doab (RD) and the Liuyuankou Irrigation System (LIS). The MIA has an arid climate with low rainfall whereas Rechna Doab and LIS both receive considerable rainfall. In all cases rainfall is clearly below potential evapotranspiration.

Australia – The Murrumbidgee Irrigation Area

The Murrumbidgee Irrigation Area (MIA) is situated in the central New South Wales region of south-east Australia. Irrigation suitability studies were undertaken along the Murrumbidgee River in the 1890s, with development taking place between 1906 and 1913. By 1914, there were 677 farms in the MIA. Water was supplied by the first major reservoir built for irrigation – Burrinjuck Dam, which was completed in 1924. Rice growing started in the MIA in 1924, although rapid development of rice areas occurred in the 1970s and 1980s. The total area for the MIA is 156,605 ha and the main agricultural products are rice, grapes and citrus. Rice is the most dominant water user with more than 32,000 ha (14 % of the total landscape) in 2000. Irrigation demand for crop production is mainly met by water drawn from the channels and less so by groundwater pumping

These inputs and outputs have been aggregated from a spatial model (750m grid) of the MIA to provide a lumped water balance for discussion in this paper. From the point of view of overall sustainability, it is necessary to consider total flow to and from the aquifer system in relation to total lateral outflow potential of the system. If there is continuous accumulation of flows and salts then the area will eventually become waterlogged and salinised. This has already happened in many of the horticultural areas, which have been under an intensive irrigation regime for 60 years and therefore needed artificial drainage. The total (lateral+ pumping) outflow of the aquifer, within the considered spatial boundaries of the system, is less than the total (vertical recharge and lateral) inflow to the aquifer. If these trends continue the groundwater pressure will rise and overall rate of soil salinisation will increase. This will result in substantial yield decline over time, environmental degradation within and outside the area, depreciation of natural capital and therefore a corresponding reduction in water use benefits from the system. For long term sustainability of this system there is a need to export salts from the area by establishing surface and subsurface drainage.

China – The Liuyuankou Irrigation System

The Liuyuankou Irrigation System (LIS) is located in Kaifeng County in the Chinese province of Henan and has been operational since 1967. The major crops are maize, rice and cotton. Eight branch channels were constructed between 1984 and 1988. Irrigation for crop production is met by water drawn from the channels as well as by groundwater pumping. During recent years irrigation conditions have become more efficient due to the improvement and maintenance of the hydraulic structures. In spite of this improved efficiency and the presence of a drainage system, the groundwater table has risen alarmingly. In the northern part of the LIS the groundwater tables are very shallow, within 1 m of the land surface. The lateral outflow of the aquifer is very small compared to the total inflow of water (within the considered spatial boundaries of the system), a significant amount of the irrigation water leaves LIS through fallow evaporation and crop transpiration.

A significant component of the overall water balance is the large lateral seepage from the Yellow River, which equates to an overall water supply of a similar order to the surface diversion. This is attributed to greater hydraulic conductivity of the aquifers and the height of the Yellow River above the surrounding plains. The lateral outflow from the aquifer, within the considered spatial boundaries of the system, is very small compared with the total vertical recharge and lateral inflow to the aquifer. The groundwater aquifer is already full and there is risk of soil salinisation if hydraulic loading due to rice is reduced as it is mainly responsible for pushing salts down through the aquifer system. Since the overall outflow is very small this area is a net salt sink and is recycling these salts through the system by groundwater pumping. This feature of the system could cause substantial yield decline in the future. There is need to change groundwater pumping to shallow watertable areas and introduce more surface water supplies in the present groundwater-dependent areas.

Pakistan – The Rechna Doab

The Rechna Doab ("land between two rivers") is the interfluvial sedimentary basin of the Chenab and Ravi rivers in Pakistan. It is one of the oldest, agriculturally-richest and most intensively-populated irrigated areas of Punjab Province. Irrigation water is pumped from the aquifer and drawn from the rivers. The gross area of Rechna Doab is 2.97 Mha, with a longitudinal extent of 403 km and a maximum width of 113 km. The area falls in the rice-wheat and sugarcane-wheat agro-climatic zones of the Province, with rice, cotton and forage crops dominating in summer, wheat and forage in winter. In some parts, sugar cane is also cultivated as an annual crop.

In the upper part of Rechna Doab a significant amount of surface water is used for irrigation. Relatively low volumes of surface water are available in the lower part of Rechna Doab, therefore crop demand is met to a larger extent by groundwater pumping. The declining groundwater levels in the lower part of Doab are reducing profitability as costs for pumping groundwater increase. In contrast to the upper Rechna Doab, there is no evaporation from fallowed soil in the lower Doab. However, salinity is a bigger problem there because the groundwater used for irrigation contains increasing concentration of salt in water leached water below the root zone of crops. Larger scale drawdown of the watertable in the lower part of Doab is also promoting lateral flow of saline groundwater from its central part.

Conclusions

The challenges faced by present day irrigated agriculture are not much different to those faced by ancient agricultural systems such as those of Mesopotamia. A close examination of the water balance components of three irrigation systems in Australia, China and Pakistan show that surface water efficiency is highest for the Murrumbidgee Irrigation Area (over 77 %) whereas surface water efficiency of Rechna Doab in Pakistan and LIS in China are less than 50 %. All these systems are dependent on direct or indirect use of groundwater. The direct shallow groundwater uptake by crops is very high in the LIS and MIA. If the direct groundwater use by crops continues in both the MIA and LIS, it will accelerate the rate of salinisation of soils. In the case of Rechna Doab more than 50% of crop water requirements are met from groundwater pumping which may result in salinity and sodicity unless adequate leaching of the root zone is not maintained. A key question for systems such as Rechna Doab and LIS is whether it is more cost effective to reduce seepage from channels or to pump groundwater.

There is a need to quantify regional water quality trends, downstream environmental impacts and the trade-off between yield reduction and direct regional groundwater use by crops in these systems. We can maintain the productive function of any of these areas by providing adequate drainage and salt export facilities, which have high energy and capital requirements. The most cost-effective option may be to increase water use efficiency and reduce negative impacts on the environment thereby reducing the associated costs of maintaining our natural capital. There is a need to radically rethink sustainability of food production, rational pricing and sharing of water and commodities to justify investment that will maintain and enhance ecosystem function within irrigated catchments. Under present operational conditions none of the three systems discussed in this paper is sustainable.

9

Irrigation Impact on Agricultural Growth and Poverty Alleviation

Though the positive impact of irrigation on agricultural intensification and increased crop yield has been very well documented by various studies, the marginal returns of irrigation versus other factor inputs such as, farm technology and other rural infrastructure development are still a controversial issue in rural development literature.

Improved information and understanding of the scale of incremental benefit of irrigation and other factor-inputs to agricultural growth and development and to poverty alleviation process has large public policy implications on setting rural development policy in a region. This has particularly more relevant in setting irrigation and agricultural investment and financing policies. This information is also important considering the recently increased global public policy priorities and thrusts on poverty reduction strategies.

In this study an attempt is made to analyse the incremental impact (input specific effects) of irrigation and other factor inputs on growth of agricultural productivity of all inputs taken together (i.e., Total Factor Productivity) and their implications on poverty alleviation in India over the last two and half decades. The Total Factor Productivity is also called as productivity of all inputs taken together, and it is different than the conventionally understood productivity measures like crops yield, water productivity or labour productivity. In addition, the study also examines the structure and relative importance of the factors that affect the over time variation in poverty and rural consumption level across the states in India. This is done by using annual time series and cross section (state) data analysis technique across 14 major states of India covering the period from 1970 to 1994, which covers more than 90 percent of the agrarian economy of India.

The overall growth and technical change in the agricultural sector has large implications on expanding the economic base and poverty alleviation process in a region. Past empirical studies have shown that ultimately the growth in productivity of all factors (TFP) in agriculture is the backbone for alleviating rural poverty in developing countries. While summarizing the previous literature on agriculture growth and poverty reduction, Mellor (2001) points out that agricultural growth has a profound impact on poverty reduction in the developing countries including the reduction of the inequity over time.

The actual level of impact of agricultural growth on poverty in fact varies by the nature, region and time period selected for the studies. Though most of the previous studies have unequivocally demonstrated that agricultural productivity growth has a profound impact on reducing poverty in Asia, the existing literature on rural poverty has failed to examine the incremental impact of each of the factor inputs on agricultural productivity growth as well as their marginal impact on poverty alleviation, and rural income enhancement.

Several irrigation impact related case studies and regional studies in India have illustrated that irrigation management has a profound role to play in the poverty alleviation process. Some of the recent aggregate level empirical studies in India have also shown that irrigation access has a positive impact on poverty reduction. However, there is no straightforward relationship shown between irrigation and poverty alleviation and the irrigation impact on poverty alleviation depends upon several other intermediate factors. Thus an improved understanding of the structure of the impact of various factors and a quantification of the marginal impacts of each of the factor input on poverty measures is important for developing efficient and effective policy instruments in poverty alleviation and rural development programs in general..

Results and Discussions

Factors Affecting Productivity of all Inputs

This study quantifies the marginal impact of irrigation and other factor inputs on the agricultural productivity of all inputs and on the two key poverty measures across the states. The empirical results show that there is no significant growth taking place in the agriculture productivity level when the level of all inputs use and their costs are taken together (in terms of economic and technical efficiency) in India over the last two decades. This productivity growth of all inputs is different than the simple crops yield or labour productivity.

This means that one percent increase in irrigated area has increased about 0.32 percent in the productivity of all inputs (TFP) in India during 1970-1994. This is a very high impact on agricultural productivity when compared to the impact of other factors such as fertilisers, HYV and road infrastructure, where the elasticity varies only from 0.04 to 0.09. The impact of rural literacy rate (percent of rural population) is positive and statistically highly significant in explaining the over time variation in agricultural productivity of all inputs (TFP). The marginal impact of rural literacy on agricultural productivity is the largest among the variables selected for the analysis. This large impact of rural education is possible considering the fact that technical change and agricultural productivity (represented by increased TFP index) and rural development are directly related to the adoption of improved technology, selection of appropriate mix of crops and inputs, and timely application of these inputs, including the farmers' ability to effectively process market and price information and farm managerial decisions. The impact of variable road infrastructure is also positive and significant which may be capturing the effects of market access in agricultural and rural development.

Factors Affecting Poverty Rate and Per Capita Rural Consumption Level

The study has also analysed the direct impact of selected factor inputs in explaining the variation in poverty measures (poverty measure in head count ratio and rural per capita consumption) across the states for 1970-1993, using the same set of factors used on analysing the agricultural productivity. The study found that the extent of rural poverty was unequivocally higher in a state with less extent of irrigation especially in the early 1970s.

Moreover, the relationship between rural poverty and irrigation has been decreasing in the recent past. This shows that the scale of poverty level was very severe in the early 1970s, with more than 60 percent of the rural population under poverty line (head-count ratio) in almost all parts of central and eastern India. Irrigation development was also very low in these regions at that time. However, the situation has improved in the early 1990s where rural poverty is mostly concentrated in states like Bihar, Orissa, and Madhya Pradesh, where irrigation development is poor as of today. While analysing the independent relationship between irrigation and rural poverty, there appears to be a strong inverse relationship between the incidence of rural poverty and percentage of gross cropped area irrigated.

Besides analysing the role of irrigation in rural poverty through regression analysis, we have also depicted the relationship between irrigation and poverty in the two graphs. The trend on variation in irrigation and the various measures of poverty, and how they have changed overtime is illustrated. The level of irrigation has increased more than double between 1960 and 1990. All the poverty measures have declined unequivocally during these periods. The Head Count Index (HCI) which measures the percentage of population below poverty line using consumption expenditures has reduced over the period of the last three decades. More importantly, the other two measures of poverty, poverty gap index (PGI) and Foster-Greer-Thorbecke (FGT) have declined at faster rate over the period of last three decades. This means all three measures of rural poverty in India have declined during the period, largely contributed by the success of irrigated agriculture over the years. As depicted through figures, the regression results also clearly demonstrate the role of irrigation in reducing the rural poverty.

The negative sign of time trend variable in poverty model, which shows overtime change on trend of poverty rate, suggests that poverty level in India has unequivocally decreased during the time period of 1970-1993. This is also supported by the positive sign of this time trend variable in the consumption model which shows over time increasing rate of per capita consumption of rural population. Among all the variables selected for analysing the poverty measures in this study, the irrigation factor has the strongest influence in explaining the reduction in poverty.

Irrigation has even a larger marginal impact on reducing the poverty than the impact of rural literacy. Whereas rural education factor was earlier found to be strongest in influencing the agricultural productivity across the states. Likewise, the increased HYV adoption and fertilizers use have also played a favourable role in reducing poverty in India, but their influence on poverty reduction is lower than the marginal incremental impact of irrigation and rural literacy. Unlike in productivity growth, the variable road infrastructure does not play any positive and favourable role in explaining the variation in rural poverty in India during the time period selected for this study.

India's Groundwater Challenge

Groundwater is a dry topic unless you happen to have a dry well. That said, groundwater management involves some of the most complex and socially challenging sets of issues facing India in the 21st century. Furthermore, how those issues are resolved will affect both the environment and the day to day life of most people living in rural and urban areas. Groundwater is an invisible resource. As a result, both the dynamics of the resource base and the services it produces are often poorly understood. This article focuses first on the broad array of environmental and social services that depend on groundwater. Key aspects of the resource base and emerging problems are presented next. The final section outlines some of the social, ethical and institutional challenges inherent in sustainable management.

Over the past 50 years, expansion of groundwater irrigation has played a lead role in food security. Yields in areas irrigated by groundwater are often substantially higher than yields in areas irrigated from surface sources. In India, for example, research indicates that yields in groundwater irrigated areas are higher by one third to one half than in areas irrigated from surface sources and as much as 70-80% of India's agricultural output may be groundwater dependent.

Higher yields from groundwater irrigated areas are due, in large part, to its ease of control and reliability. Early studies indicated that water control alone can reduce the gap between potential and actual yields by about 20%. This translates into substantial benefits. Reliability is even more important. Groundwater is a key buffer against drought and normal variations in rainfall. Overall, increased yields from groundwater irrigated areas have translated into substantially higher yields and are thus a major factor in food production at the regional and national levels. Furthermore, some of the most important food security benefits related to groundwater lie at the level of individual farmers. The vulnerability to natural hazards of different groups in society, including those that threaten food security, can be explained by their access to networks of key productive and social resources. For rural populations, groundwater is among the most important of these resources.

Households with access to key resources are able to build support systems that reduce their vulnerability to natural hazards. Groundwater irrigation reduces the risk that investment in labour, seed, fertilisers, pesticides and other inputs will be lost due to drought or the variability of precipitation in normal years while higher yields enable households to generate surpluses. As a result, households with access to groundwater tend to have higher levels of savings and are able to make investments in other productive resources or activities.

When drought strikes or there is a gap in rainfall these households have a dual advantage. First, they are far less likely to suffer losses than those without access to groundwater. Second, even if 'the well runs dry', households that own wells have often been able to save cash or food and invest in alternative sources of income. As a result, they have assets that can carry them through periods of scarcity or crisis.

Groundwater is a highly important source of domestic water supply. In India, roughly 80% of rural water supply for domestic uses is met from groundwater. The importance of potable drinking water is clear. As in the case of other uses, however, this is only a portion of the value of groundwater as a source of domestic supply. Wells in villages and towns free people, particularly women, from long daily walks to fetch water from springs or rivers for livestock and domestic uses. This frees time and labour for other activities. Furthermore, since water no longer has to be carried over long distances, more is often used. This can have major health benefits. In addition, because of the filtering nature of the soil and frequent long residence time underground, groundwater is commonly much cleaner than surface sources.

Groundwater is a key resource for poverty alleviation and economic development. Evidence indicates that improved water sources generate many positive externalities in the overall household micro-economy. In areas dependent on irrigated agriculture, the reliability of groundwater sources and the high crop yields generally achieved as a result often enable farmers with small landholdings to increase income. In India small and marginal farmers (those having less than 2 hectares) own 29% of the agricultural area. Their share in net area irrigated by wells is, however, 38.1 % and they account for 35.3 % of the tubewells fitted with electric pump sets.

Thus, in relation to operational area, small and marginal farmers tend to have proportionally more irrigated land than larger farmers. With productivity on irrigated lands being much higher than that on non-irrigated tracts, better access to irrigation for small and marginal farmers can significantly reduce poverty.

The positive economic impact of groundwater development extends beyond well owners. Access to groundwater stabilises the demand for associated inputs, and leads to the spread of support services for pumps and wells, creating a base for small scale rural industries.

Furthermore, the spread of groundwater irrigation can increase demand for labour. In India, for example, labour accounts for approximately 44 % of the cost of installing a well and the additional indirect employment created on every hectare of irrigated land through increased agricultural activity is approximately 45 days per hectare. Therefore, the expansion of groundwater irrigation has significant ripple effects, creating employment throughout rural economies.

The equity impacts of groundwater development for irrigation are, however, not all positive. Modern tubewell and drilling technology tends to be capital intensive. As a result, early exploiters of groundwater have typically been large farmers who produce surpluses for the market Small holders growing subsistence crops often depend on supplementary groundwater irrigation using a variety of man and animal-driven water lifting devices from shallow, open wells. The expansion of energised pumping technologies tends to draw water levels down, driving shallow wells and muscle-driven devices out of business. This was, for instance, the case throughout the Gangetic basin and other parts of India during the 1960s.

As water levels began to decline, some state governments in India attempted to implement administrative regulations, such as selective credit controls, restrictions on electricity connections and siting and licensing rules. These regulations did not affect landowners who were able to install tubewells early, but limited the entry of latecomers – particularly the resource poor who depended on credit or access to subsidised electricity in order to afford the capital cost of operating and installing pumps. In addition, the economically and politically powerful were generally able to bypass regulations via 'adjustments' with officials or by depending on their own financial resources for well construction and operation.

Equity considerations are generally a major point of tension as management needs emerge. Rapid unrestricted development of groundwater has reduced poverty by giving the poor access to a key resource for production. This same pattern of unrestricted development, however, is the primary cause of over-extraction and quality problems now emerging in many parts of the world. As groundwater problems grow, marginal populations are often the first affected. Water level declines, for example, have the largest economic impact on individuals who are unable to afford deeper wells – i.e. the poor. The poor are also the least well positioned to protect their interests if groundwater extraction must be reduced. Restrictions on new wells tend to affect them much more than wealthy communities where wells were installed much earlier. Wealthy individuals and communities are also often able to work around these and other types of management restrictions while poorer communities (who generally lack political as well as economic leverage) have less ability to do so. In sum, there is an inherent tension between equitable access to groundwater for all sections of society and sustainable management of the resource base.

The array of environmental services or values dependent on groundwater are often poorly understood. Environmental concerns related to groundwater generally focus on the impacts of pollution and quality degradation on human uses, particularly domestic supply. Development impacts on the groundwater environment are, however, different from the numerous environmental services provided by groundwater resources in their natural state.

What are some of the most important environmental services provided by groundwater? The following list, while not exhaustive, illustrates the degree to which many environmental values are dependent on groundwater:

- *Catchment base flow* derived from groundwater discharge is perhaps the most evident environmental value associated with groundwater. In many areas springs and the dry season flow in rivers depend heavily on groundwater.
- *Instream fisheries and aquatic ecosystems*: Instream flows are critical for the maintenance of fisheries and aquatic ecosystems. As previously noted, groundwater contributions can be a dominant source of water for instream flows, particularly during droughts and dry seasons.
- *Inland wetlands*: Wetlands are some of the most productive and biologically diverse inland ecosystems. In many, if not most, cases water availability in wetlands depends on high groundwater levels.
- *Surface vegetation*: Groundwater levels directly influence many vegetation communities. Phreatophytes, plants that derive a major portion of their water needs from saturated soils, can be the dominant vegetative species in ecosystems where groundwater levels are shallow. They often form critical wildlife habitat and may serve as important sources of food, fuel and timber.

In many areas the environmental values dependent on groundwater conditions are closely intertwined with a broad array of human use patterns. Groundwater is an integral part of linked hydrologic, ecologic and human use systems. Changes in surface water use, groundwater use or vegetation can send ripple effects throughout these inter linked systems – often with effects that are difficult to predict.

Throughout much of South Asia, groundwater development has proceeded at an exponential pace over the past four decades. In India, the number of shallow tubewells doubled roughly every 3.7 years between 1951 and 1991. Rapid development has engendered its own set of issues.

In many arid and hard rock zones, overdraft and associated quality problems are increasingly emerging. Although the area currently affected by groundwater overdraft may be limited, blocks classified as dark or critical increased at a continuous rate of 5.5% over the period 1984-85 to 1992-93. If continued at this rate, the number would double every 12.5 years. This implies that by the year 2017-18 (25 years from 1992-93), roughly 1532 blocks or 36% of the 4248 blocks in the listed states would be dark or critical.

Overdraft is, however, only a fraction of the management challenge associated with groundwater. Large areas, particularly in the command of surface irrigation systems suffer from waterlogging and associated salinity or alkalinity problems. Furthermore, development impacts on the environment and non-agricultural users can be major even where overdraft or waterlogging are absent. Seasonal watertable fluctuations can affect shallow wells, low season flows in surface streams, and pollution loads. The impact of this on drinking water availability, the poor and the environment can be major.

Furthermore, it is important to recognize that *overdraft and water level declines typically affect the sustainability of uses that are dependent on groundwater long before the resource base itself is threatened with physical exhaustion.* Many uses and environmental values *depend on the depth to water* – not the volume theoretically available. In the case of the Ganges basin, for example, water level declines would exclude the poor from access to groundwater (due to the cost of increasing well depth) and would reduce base flows in streams long before the aquifer would face any threat of depletion. The Ganges basin contains, in some locations, over 20 thousand feet of saturated sediment. Dewatering of only the top few tens of feet would, however, have tremendous economic and environmental impacts.

Beyond overdraft and water level declines lie the questions of water quality and pollution. Pollution or quality declines can cause reductions in water availability that are far less reversible than overdraft. Non-point source pollution from agriculture and other sources combined with point source pollution represents a major management challenge. Furthermore, not all quality problems are human induced. Probably the most extensive case of arsenic poisoning from groundwater is that of Bangladesh and West Bengal. Arsenic occurs naturally in the ground waters abstracted from the alluvial deltaic sediments of the Ganges-Brahmaputra-Meghna river systems and an area around 75,000 km2 is thought to be affected by groundwater with high arsenic concentrations.

Despite the widespread nature of emerging problems, how extensive groundwater mining and pollution problems really are remains uncertain. Official statistics on the number of blocks where extraction is approaching or exceeds recharge may be misleading since great uncertainty exists over the reliability of published extraction and recharge estimates. Furthermore, even the basic water table measurements on which these estimates rest may, in some cases, be open to question. Uncertainty is also inherent in relation to the extent of pollution. Clearly pollution loads have increased substantially over recent decades with increases in the use of agricultural chemicals, industrial discharges and urban waste. At the same time, no comprehensive data sets are available that would allow identification of the extent and distribution of different pollutants. Quality problems are also often identified 'after the fact'. Again the arsenic case is illustrative. This was only recognized as a widespread problem when large scale cases of arsenic poisoning began to appear.

A critical challenge in interpreting both quantity and quality problems relates to understanding of the resource base and its dynamics. Many individuals, groundwater professionals included, conceptualize groundwater as flowing smoothly through the earth with rapid recharge from rainfall and relatively uniform water quality. In reality, however, complex rock formations and differential recharge rates result in far more complicated dynamics. These, in turn, greatly complicate understanding of resource conditions. Two examples are illustrative. In Rajasthan, local communities often suggest that groundwater overdraft problems are alleviated whenever rainfall is above normal. In many cases, however, the water they pump has been underground for hundreds and, in some cases, as much as 20,000 years. Recharge dynamics depend on the permeability of soils and flow patterns are often only weakly related to short term fluctuations in precipitation. Similarly, major quality problems can depend heavily on very localized conditions. Arsenic, for example, is preferentially mobilized under reducing conditions. As a result, the amount of arsenic encountered in water from a given well can depend on the amount of organic matter available precisely where the well was drilled. A well that happens to pass near a tree trunk buried deep underground may have substantially more arsenic than a well drilled only a short distance away.

Debates over the need for management are often clouded by uncertainty over the extent and nature of emerging problems. At the same time, delays in initiating management can have irreversible implications. Once polluted, groundwater aquifers are often impossible to remediate. Pollutants can become attached to the aquifer matrix (the sand, soil and rocks) and serve as a continuous source of contamination. Overdraft can lead to aquifer compaction (reduction of pore spaces) making recharge impossible. Even if recharge is technically feasible and sufficient water is available, the low flow rates characteristic of many fine grained aquifers can make the process so slow that little can be done on a human time scale.

As a result, overdraft is often equivalent to mining a resource such as oil that can not be replenished. Even experts often have difficulty identifying the emergence of irreversible problems until after the fact. Managing uncertainty is thus a major part of any groundwater management equation. Groundwater is an invisible, poorly understood resource, yet one that is critical to a wide variety of social, economic and environmental services. Pollution and declining water levels represent direct threats to the sustainability of environmental, domestic, agricultural and industrial uses dependent on groundwater. In addition, as demands grow and the limits of sustainable extraction become evident, competition between agricultural and other users is increasing rapidly.

This can generate competitive extraction between individuals. Each person extracts as much groundwater as possible in order to capture benefits for themselves before the resource is exhausted. The net result can be a spiral of growing demands and decreasing availability. Competition is, thus, a critical social issue that must be addressed in order to manage groundwater on a sustainable basis.

Technical options for resolving competition over groundwater are generally limited. In most cases communities and groups affected by groundwater overdraft advocate aquifer recharge. Throughout India the real viability of this option is often limited. In much of Gujarat, for example, estimates indicate that increases in recharge could only reduce overdraft by 10% and in many other areas little unutilized surface water is available that could be recharged. Even where water is available, the timing and distribution of rainfall limit the ability to recharge aquifers. Recharge is a slow process limited by the infiltration rates of water through soil and underlying formations. Precipitation, in contrast, is often highly seasonal and occurs in brief bursts little of which can be stored for recharge.

Beyond the technical limitations in recharging aquifers, three concepts are central to understanding the social challenges inherent in resolving competition over groundwater resources. These are:

- *Interdependency*: Groundwater is a lynchpin that links and creates points of interdependency between agricultural, environmental and economic systems. Environmental values, access for the poor to water, food security during drought years and the economic viability of different crops may, for example, all depend on the maintenance of specific water table or water quality conditions. These conditions are, in turn, often dependent on water use patterns. Recharge from 'inefficient' surface irrigation systems, for example, often helps to maintain high groundwater levels and, through that, base flows in rivers, groundwater access for the poor and so on.
- *The Public Good Nature of Many Groundwater Services*: Individual users can only capture the 'extractive' values associated with groundwater, i.e. products produced by pumping and using it in a specific application. This extractive value does not, however, reflect the environmental, drought buffer and other services produced by groundwater when it is left in the aquifer. These services are *public goods* – while individuals may benefit from them, the conditions depend on the cumulative actions of all users. There are strong economic incentives for individuals to overpump groundwater or ignore their contribution to pollution of an aquifer. This leads to chronic undervaluation of groundwater when it is sold in water markets or analysed using standard economic approaches.
- *Scale*: In most cases, groundwater cannot be managed at a very local scale. Aquifers generally extend under regions encompassing anywhere from tens to thousands of villages. As a result, aquifer dynamics limit the impact individuals or villages can have on groundwater conditions. At the same time, approaches to management implemented through state agencies are often difficult to adapt in ways that reflect local or regional variations in groundwater conditions and use.

In some parts of the world, notably the United States, the approach to groundwater management hinge on a combination of private rights and regulatory mechanisms. This type of approach attempts to resolve competition by providing a measure of security for all users by specifying individual use rights and using regulation to address public good aspects. In some cases groundwater rights are established that specify the volume individuals are allowed to pump, the types of uses permitted and whether or not the water can be sold to other users. Water markets are widely advocated as a mechanism for ensuring water is allocated to the highest value uses wherever transferable water rights have been established.

Approaches based on regulation and the establishment of individual water rights and water markets face tremendous challenges in India. On a purely practical level, how groundwater rights could be established, monitored and enforced given the millions of wells and conditions in rural India, is far from clear. Regulation would also be difficult and probably highly inequitable in practice. A model bill for groundwater regulation was initially circulated by the Central Ground Water Board in the early 1970s. It proposes a highly centralized system of regulation by state agencies. Modified versions of this have now been adopted in a few locations in India but little implementation has actually occurred.

Beyond the practical limitations of approaches based on individual rights and regulation by the state, however, it is important to recognize the inherently incomplete nature of approaches based on rights, regulation and water markets. It is difficult to define water rights (whether allocated to individuals or communities) in ways that capture the interdependent and public good nature of the services produced by groundwater. When rights are transferable through market mechanisms, the values reflected in the market still tend to be the direct use values rather than the public goods. When attempts are made to regulate groundwater use and transfers in ways that protect public goods approaches rapidly become complex and inflexible – unable to respond to the diverse conditions encountered at local levels or to the dynamic nature of conditions. Because of the above limitations, approaches to groundwater management need to reflect the political nature of such decision making in order to be effective. The public good nature of groundwater services and inter linked and often poorly understood character of systems dependent on groundwater tends to generate substantial debate over management approaches and objectives. How this competition is resolved generally depends on the ability of different groups to first understand the nature of emerging problems and management options and second to insert their views into the decision making process that determines management actions.

Access to information, economic power and the ability to organize greatly influence the ability of different groups to effectively engage in this type of dialogue. Under current conditions, groundwater management decisions are effectively based on economic power – the ability of individuals to afford the costs of deepening their own wells and keep on pumping. Resolving this probably requires approaches based on balance of power concepts and explicit recognition of the political nature of management needs.

Key points of leverage that may encourage the development of effective groundwater management systems could include:

- improved access to information;
- legal standing for groups and individuals to force protection of public interest values; and
- the creation of management organizations capable of functioning at an intermediate scale (i.e. between the village and the state).

Irrigation in India

Except in southeastern India, which receives most of its rain from the northeast monsoon in October and November, dryland cultivators place their hopes for a harvest on the southwest monsoon, which usually reaches India in early June and by mid-July has extended to the entire country. There are great variations in the average amount of rainfall received by the various regions—from too much for most crops in the eastern Himalayas to never enough in Rajasthan. Season-to-season variations in rainfall are also great. The consequence is bumper harvests in some seasons, crop-searing drought in others. Therefore, the importance of irrigation in India cannot be overemphasized.

Irrigation i India has been a high priority in economic development since 1951; more than 50 percent of all public expenditures on agriculture have been spent on irrigation alone. The land area under irrigation expanded from 22.6 million hectares in FY 1950 to 59 million hectares in FY 1990, an increase of 161 percent in four decades. This increase was about 33 percent of the estimated potential. The overall strategy has been to concentrate public investments in surface systems, such as large dams, long canals, and other large-scale works requiring huge outlays of capital over a period of years, and in deep-well projects that also involve large capital outlays. Shallow-well schemes and small surface-water projects, mainly ponds (called tanks in India), have been supported by government credit but were otherwise installed and operated by private entrepreneurs. Roughly 42 percent of the net irrigated area in FY 1990 was from surface water sources. Tanks, step wells, and tube wells provided another 51 percent; the rest came from other sources.

Between 1951 and 1990, nearly 1,350 large-and medium-sized irrigation works were started, and about 850 were completed. The most ambitious of these projects was the Indira Gandhi Canal, with an anticipated completion date of close to 1999. When completed, the Indira Gandhi Canal will be the world's longest irrigation canal. Beginning at the Hairke Barrage, a few kilometers below the confluence of the Sutlej and Beas rivers in western Punjab, it will run south-southwest for 650 kilometers, terminating deep in Rajasthan near Jaisalmer, close to the border with Pakistan. A dramatic change already had taken place in this hot and inhospitable wasteland by the late 1980s. As a result, desert dwellers switched from raising goats and sheep to raising wheat, and outsiders flocked in to purchase six-hectare plots for the equivalent of US$3,000.

Progress in irrigation has not been without problems. In India, arge dams and long canals are costly and also highly visible indicators of progress; the political pressure to launch such projects was frequently irresistible. But because funds and technical expertise were in short supply, many projects moved forward at a slow pace. The Indira Gandhi Canal project is a leading example. And the central government's transfer of huge amounts of water from Punjab to Haryana and Rajasthan, frequently cited as a source of grievance by Sikhs in Punjab, contributed to the civil unrest in Punjab during the 1980s and early 1990s.

Problems also have arisen as ground water supplies used for irrigation face depletion. Drawing water off from one area to irrigate another often leads to increased salinity in the supply area with resultant effects on crop production there. Some areas receiving water through irrigation are poorly managed or inadequately designed; the result often is too much water and water-logged fields incapable of production. To alleviate this problem, more emphasis is being placed on using irrigation water to spray fields rather than allowing it to flow through ditches. Furthermore, charges of corruption and mismanagement have been levied against government-operated facilities. Cases of bribery, maldistribution of water, and carelessness are frequently raised in the media.

Another major problem has been the displacement of thousands of people, usually poor people, by large hydroelectric projects. Critics also claim that the projects are damaging to the ecology. Smaller projects and such traditional methods for irrigation as tanks and wells are seen as having less serious impact. In the late 1980s and early 1990s, the debate between large-scale versus small-scale projects came to the fore because of the US$3 billion Sardar Sarovar project on the Narmada River. Sardar Sarovar, as conceived, was one of the world's largest hydroelectric and irrigation projects. Some 37,000 hectares of land in Madhya Pradesh, Gujarat, and Maharashtra were slated to be submerged following the construction of some 3,000 dams, 75,000 kilometers of canals, and an electric power generating capacity of 1,450 megawatts of power per year. Included among the 3,000 dams was the proposed 160-meter-high Sardar Sarovar Dam. In 1985 the World Bank agreed to loan US$450 million for the project. Environmentalists in India and abroad, however, argued that the project was ecologically undesirable. In the face of this strong protest, the World Bank appointed a two-member team in 1991 to review the project. Despite a negative review of the environmental impact by the team, World Bank funding and the project continued. By 1993, however, in the face of continued international protest as well as opposition and a call for a satyagraha by villages in the affected areas, the central government cancelled the dam project loan. Work on the Sardar Sarovar project continues, however, with funds provided by the central government and the governments of the three states involved.

Although India had the second largest irrigated area in the world, the area under assured irrigation or with at least minimal drainage is inadequate. The irrigation potential estimated to have been created by the early 1990s was about 82.8 million hectares. This amount includes the gross irrigated area plus the potential for double cropping provided by irrigation. There was a cumulative gap in irrigated land use of about 8.6 million hectares until FY 1990, by which time the gap had decreased through improved land management.

Irrigation Water Requirements

The assessment of the irrigation potential, based on soil and water resources, can only be done by simultaneously assessing the irrigation water requirements (IWR).

Net irrigation water requirement (NIWR) is the quantity of water necessary for crop growth. It is expressed in millimetres per year or in m^3/ha per year (1 mm = 10 m^3/ha). It depends on the cropping pattern and the climate. Information on irrigation efficiency is necessary to be able to transform NIWR into gross irrigation water requirement (GIWR), which is the quantity of water to be applied in reality, taking into account water losses. Multiplying GIWR by the area that is suitable for irrigation gives the total water requirement for that area. In this study water requirements are expressed in km^3/year. Calculations of irrigation water requirements are done while preparing national water master plans or irrigation projects. Useful information was obtained from a number of country studies available from AQUASTAT, but the information was based on many different approaches. For the purpose of this study the need was felt to develop a method of computing irrigation water requirements for the whole continent in a systematic way. In order to be able to do this at the scale of the continent, assumptions have to be made on the definition of areas to be considered homogeneous in terms of rainfall, potential evapotranspiration, cropping pattern, cropping intensity and irrigation efficiency.

Methodology

For the calculation of irrigation water requirements the following steps have been followed:

- Delineation of major irrigation cropping pattern zones. These zones are considered homogeneous in terms of types of irrigated crops grown, crop calendar, cropping intensity and gross irrigation efficiency. Represented on the map of Africa, they should be viewed as regions where some homogeneity can be found in terms of irrigated crops. The cropping pattern proposed for the zone should be viewed as representative of an 'average' rather than a 'typical' irrigation scheme.
- Definition of the area of influence of the climate stations (in GIS) and quality check on the climate data.
- Combination of the irrigation cropping pattern zones with the climate stations' zones (in GIS) to obtain basic mapping units.
- Calculation of net and gross irrigation water requirements for different scenarios.
- Comparison with existing data and final adjustment.

Delineation of Irrigation Cropping Pattern Zones

The criteria used for the delineation of the irrigation cropping pattern zones were, in order of decreasing importance: distribution of irrigated crops, average rainfall trends and patterns, topographic gradients, presence of large river valleys (Nile, Niger, Senegal), presence of extensive wetlands (the Sudd in Sudan), population pressure, technological differences and crop calendar above and below the equator (Zaire). The starting point was the type of irrigated crops currently grown in Africa. This resulted in 18 zones. From these zones, sub-zones showing a different cropping intensity or a different crop calendar were defied. This resulted in a total of 24 irrigation pattern zones, which are considered to be homogeneous for:

- crops currently grown;
- crop calendar;
- cropping intensity.

Only the main crops currently grown, those occupying at least 85% of the irrigated area, were considered. Land occupation of the remaining 15 % by secondary crops was assigned to the main crops. An 'average' typical monthly crop calendar was assigned to each zone, based on work done by FAO's global information and early warning system, and on information from the reference library of FAO's agro-meteorology group, AQUASTAT and, for eastern Africa, from the IGADD crop production system zones inventory.

For each crop the actual cropping intensity was derived from national crop production and land use figures extracted from the FAO AGROSTAT and AQUASTAT databases. It ranges from 100 to 200%, according to the crop calendar. The cropping intensity to be used in this study of irrigation potential ('potential' scenario) was generally estimated by increasing current values by 10 to 20%, but it was assumed that because of market limitations the current high intensity (in relative terms) of vegetables in certain parts of the continent would not be found in the potential scenario. Therefore, intensities of cereal crops are higher in the potential scenario than in the actual situation.

Definition of the Climate Stations' area of Influence

The climate data from the FAOCLIM cd-rom were used, as this was the most up to date climate database available. This data set includes long term average rainfall and reference potential evapotranspiration (ET_0) data for 1025 stations throughout Africa. ET_0 was calculated by the Penman-Monteith method.

To obtain a spatial coverage of climate data (P. ET_0) over the continent, each station was assigned an area of influence using the Thiessen polygons method. This method assigns an area of 'nearest vicinity' to each climate station. Gives an indication of the density of the stations over the continent. As expected, the desert areas in northern and southern Africa are much less well covered than the rest of the continent. The rainfall data were compared with raster maps prepared by the Australian National University and corrected where necessary.

Combination of Cropping Pattern Zones with the Climate Stations

In ArcInfo, the 24 cropping pattern zones and the 1025 climate station data were merged. This resulted in 1437 basic map features, homogeneous in irrigation cropping characteristics and climate. All further calculations were carried out on these 1437 basic mapping units.

Results

Summarizes the figures for each of the 84 zones. NIWR and GIWR for the potential scenario with effective rainfall were further combined with the 136 basic units of this study to obtain individual NIWR and GIWR for each of these units through GIS. The results have been compared with figures available from country studies (national water master plans, projects, etc.). The comparison shows that the methodology yields relatively accurate regional estimates of IWR that are suitable for the present study. Discrepancies with country studies find their origin mostly in the assumptions made on cropping pattern, cropping intensity and irrigation efficiency.

The influence of *cropping pattern zones* on the quality of the output is of prime importance. Important differences in irrigation water requirements in adjacent zones is one of the consequences of this approach. For instance, in Burkina Faso, areas located north of the 1000 mm annual rainfall line have a gross potential water requirement of 500 mm per year, while areas located just south of this line need more than 2800 mm per year. This artificial break is due to the choice of the irrigated cropping pattern zones, where it was decided that no rice was cultivated under 1000 mm of rainfall per year. Within the cropping pattern zones, the boundaries of irrigation water requirement zones follow rainfall trends.

The differences in irrigation water requirements between adjacent zones is directly related to the *density of the climate stations' network*. In low-density areas, such as the Sahara and southern Africa, differences in IWR between adjacent zones are high (up to 600 mm/ year gross requirements) as the low station density does not allow the delineation of HIWR zones with smaller differences. A high density of the station network in the rest of the continent, in combination with rainfall raster maps, has resulted in differences of a maximum of 200 mm/year gross requirements between adjacent zones.

Irrigation Cropping Pattern Zones. List of Cropping Pattern Zones.

1. Mediterranean coastal zone
2. Saharan oases

3a Semi-arid to arid savannas in West-East Africa
3b Semi-arid/arid savanna (Somalia-Kenya-Southern Sudan)
4a Rice: Niger/Senegal rivers
4b Rice: Gulf of Guinea
4c Rice: Southern Sudan
4d Rice: Madagascar tropical lowland
4e Rice: Madagascar highland
5 Egyptian Nile and delta
6 Ethiopian highlands
7 Sudanese Nile area
8 Shebelli-Juba river area in Somalia
9 Rwanda-Burundi-Southern Uganda highlands
10 Southern Kenya-Northern Tanzania
11 Malawi-Mozambique-Southern Tanzania
12a West and Central African humid areas above the equator
12b Central African areas below the equator
13 River effluents on Angola-Namibia-Botswana border
14 South Africa-Namibia-Botswana desert and steppe
15 Zimbabwe highland
16 South Africa-Lesotho-Swaziland
17 Awash river area in Ethiopia
18 All islands (Comoros-Mauritius-Seychelles-Cape Verde).

Environmental Considerations in Irrigation Development

Irrigation has contributed significantly to poverty alleviation, food security, and improving the quality of life for rural populations. However, the sustainability of irrigated agriculture is being questioned, both economically and environmentally. The increased dependence on irrigation has not been without its negative environmental effects.

Inadequate attention to factors other than the technical engineering and projected economic implications of large-scale irrigation or drainage schemes in Africa has all too frequently led to great difficulties. Decisions to embark on these costly projects have often been made in the absence of sound objective assessments of their environmental and social implications. Major capital intensive water engineering schemes have been proposed without a proper evaluation of their environmental impact and without realistic assessments of the true costs and benefits that are likely to result.

The sustainability of irrigation projects depends on the taking into consideration of environmental effects as well as on the availability of funds for the maintenance of the implemented schemes. Negative environmental impacts could have a serious effect on the investments in the irrigation sector. Adequate maintenance funds should be provided to the implementing organizations to carry out both regular and emergency maintenance. It is essential that irrigation projects be planned and managed in the context of overall river basin and regional development plans, including both the upland catchment areas and the catchment areas downstream.

Potential Environmental Impacts of Irrigation Development

The expansion and intensification of agriculture made possible by irrigation has the potential for causing: increased erosion; pollution of surface water and groundwater from agricultural biocides; deterioration of water quality; increased nutrient levels in the irrigation and drainage water resulting in algal blooms, proliferation of aquatic weeds and eutrophication in irrigation canals and downstream waterways. Poor water quality below an irrigation project may render the water unfit for other users, harm aquatic species and, because of high nutrient content, result in aquatic weed growth that obstructs waterways and has health, navigation and ecological consequences. Elimination of dry season die-back and the creation of a more humid microclimate may result in an increase of agricultural pests an plant diseases. Large irrigation projects which impound or divert river water have the potential to cause major environmental disturbances, resulting from changes in the hydrology and limnology of river basins. Reducing the river flow changes flood plain land use and ecology and can cause salt water intrusion in the river and into the groundwater of adjacent lands. Diversion of water through irrigation further reduces the water supply for downstream users, including municipalities, industries and agriculture. A reduction in river base flow also decreases the dilution of municipal and industrial wastes added downstream, posing pollution and health hazards.

The potential direct negative environmental impacts of the use of groundwater for irrigation arise from over-extraction (withdrawing water in excess of the recharge rate). This can result in the lowering of the water table, land subsidence, decreased water quality and saltwater intrusion in coastal areas. Upstream land uses affect the quality of water entering the irrigation area, particularly the sediment content (for example from agriculture-induced erosion) and chemical composition (for example from agricultural and industrial pollutants). Use of river water with a large sediment load may result in canal clogging. The potential negative environmental impacts of most large irrigation projects described more in detail below include: waterlogging and salinization of soils, increased incidence of water-borne and water-related diseases, possible negative impacts of dams and reservoirs, problems of resettlement or changes in the lifestyle of local populations.

Waterlogging and Salinization

About 2 to 3 million ha are going out of production worldwide each year due to salinity problems. On irrigated land salinization is the major cause of land being lost to production and is one of the most prolific adverse environmental impacts associated with irrigation. However, very limited research has yet been conducted to quantify the economic impact of irrigation induced salinization. Quantitative measurements have generally been limited to the amount of land affected or abandoned. Estimates of the area affected have ranged from 10 to 48% of worldwide total irrigated area. Especially the arid and semi-arid areas have extensive salinity problems.

Waterlogging and salinization of soils are common problems associated with surface irrigation. Waterlogging results primarily from inadequate drainage and over-irrigation and, to a lesser extent, from seepage from canals and ditches. Waterlogging concentrates salts, drawn up from lower in the soil profile, in the plants' rooting zone. Alkalization, the build-up of sodium in soils, is a particularly detrimental form of salinization which is difficult to rectify. Irrigation-induced salinity can arise as a result of the use of any irrigation water, irrigation of saline soils, and rising levels of saline groundwater combined with inadequate leaching. When surface water or groundwater containing mineral salts is used for irrigating crops, salts are carried out into the root zone. In the process of evapotranspiration, the salt is left behind in the soil, since the amount taken up by plants and removed at harvest is quite negligible. The more arid the region, the larger is the quantity of irrigation water and, consequently, the salts applied, and the smaller is the quantity of rainfall that is available to leach away the accumulating salts. Excess salinity within the root zone reduces plant growth due to increasing energy that the plant must expend to acquire water from the soil. The tolerance of crops to salinity is variable: clover and rice are more sensitive to salts than barley and wheat. Comprehensive studies of farm-level effects of irrigation-induced salinity indicate that the yields of paddy and wheat are around 50% lower on the degraded soils and net incomes in salt-affected lands are around 85% lower than the unaffected land.

Irrigation-related salinity has adverse effects not only on the production areas, but also on areas and people downstream. The rivers, particularly in arid zones tend to become progressively more saline from their headwaters to their mouths. The aquifers interrelated with the river are highly saline and the salts discharged to the river system from saline aquifers adversely affect downstream water users, particularly irrigated agriculture and, in some special cases, wildlife. Many of the soil-related problems could be minimised by installing adequate drainage systems. In Egypt, for example, the installation of drainage systems effectively reduced soil salinity. The average yield for wheat increased from 1 ton/ha before drainage to about 2.4 tons/ha. Similarly, the yield for maize increased from 2.4 tons/ha to 3.6 tons/ha after drainage infrastructure was completed. Drainage is a critical element of irrigation projects, that however still too often is poorly planned and managed. Waterlogging can also be reduced or minimized, in some cases, by using micro-irrigation which applies water more precisely and can more easily limit quantities to no more than the crops needs.

Water-borne and Water-related Diseases

Water-borne or water-related diseases are commonly associated with the introduction of irrigation. The diseases most directly linked with irrigation are malaria, bilharzia (schistosomiasis) and river blindness (onchocerciasis), whose vectors proliferate in the irrigation waters. Other irrigation-related health risks include those associated with increased use of agrochemicals, deterioration of water quality, and increased population pressure in the area. The reuse of wastewater for irrigation has the potential, depending on the extent of treatment, of transmitting communicable diseases. The population groups at risk include agricultural workers, consumers of crops and meat from the wastewater-irrigated fields, and people living nearby. Sprinkler irrigation poses an additional risk through the potential dispersal of pathogens through the air.

The risk that one or more of the above diseases is introduced or has an increased impact is most likely in irrigation schemes where:

- soil drainage is poor, drainage canals are either absent, badly designed and/or maintained;
- rice or sugar cane is cultivated;
- night storage reservoirs are constructed;
- borrow pits are left with stagnant water;
- canals are unlined and have unchecked vegetation growth.

Malaria

Malaria is by far the most important disease, both in terms of the number of people annually infected, and whose quality of life and working capacity are reduced, and in terms of deaths. Worldwide, some 2000 million people live in areas where they are at risk of contracting malaria. The total number of people infected with malaria is estimated at 100 to 200 million with between 1 and 2 million deaths per year, with almost 90% of the cases in Africa. Drug treatment has become difficult recently because the parasite has become resistant to certain drugs that have been used for a long time in many parts of the world. Interruption of disease transmission chemicals for the control of the vector, the mosquito, has become less effective because some mosquito vector species have become resistant to previously effective insecticides and some insecticides have been banned for environmental reasons.

Bilharzia

Bilharzia is almost as widespread as malaria, but rarely causes immediate death. An estimated 200 million people are infected and the transmission occurs in some 74 countries. The infection is particularly common in children who play in water inhabited by the snail intermediate host.

Severe infection in childhood leads to long-term damage to bladder, kidneys and liver, which may cause death many years after the original infection. Infection at any age may make people feel unwell and reduce working capacity.

Bilharzia is an infection caused by parasitic worms or blood flukes of certain species of the genus *Schistosoma*. Adult parasites live in the blood of mammals, but their life cycle requires a phase of asexual multiplication within a freshwater snail host. The flukes infect humans who enter their exposed skin in water, usually through swimming, bathing or wading. There exists either urinary or intestinal schistosomiasis. The type and extent of health complications associated with schistosomiasis appear to vary with species and strain of parasite and by the characteristics of the human population.

As shown in some examples below, water resources development projects may make schistosomiasis worse, and that there are serious public health consequences in many cases. Specifically, WHO stated that in areas endemic for schistosomiasis, water resources development projects should have schistosomiasis prevention and control built into programme design and implementation. Furthermore, even in cases where irrigation does not increase schistosomiasis infection rates, careful studies should be made of snail species and of existing patterns of schistosomiasis transmission. Further, a percentage of investment and operating funds should be allocated for appropriate water supply and sanitation and for health care to treat local populations for any water-related or other ailments associated with the project.

Effects of Large Dams and Reservoirs on the Prevalence of Bilharzia in Africa

In Africa, in recent decades cautionary warnings have accompanied many irrigation and dam projects regarding the likely impact of increased schistosomiasis transmission. In some cases, these prophecies came true, including the Volta Lake project in Ghana in the 1960s and the Kainji Lake project in Nigeria around 1970. Most recently, research appears to confirm the prediction, that constructing a dam across the mouth of the Senegal River would lead to a surge in schistosomiasis transmission.

In part of Upper Egypt, schistosomiasis prevalence is said to have increased from 6 to 60% three years after the Aswan low dam was completed in the early 1930s. Following the construction of the Sennar dam in Sudan in 1926 and the Gezira scheme which followed, schistosomiasis spread. At the Arusha Chini irrigation project in Tanzania, research reported the prevalence to be 85 % in 1962. On the southwestern shore of Lake Volta in Ghana, prevalence is reported to have reached 90% two years after the lake was filled in 1966.

In Zambia, prevalence of schistosomiasis around Lake Kariba in 1968, 10 years after the Zambezi river was dammed, was 15% in adults and 70% in children. At Kainji Lake in Nigeria, prevalence increased from 30% to 45%.

In Ethiopia, the two Koka dams in the Awash Valley present a case of the different effects a dam has on the two different forms of schistosomiasis through the reservoir, downstream hydrology, and irrigation. It was reported that these two dams, built in 1958 and 1964, appeared to have no effect on urinary schistosomiasis transmission through their reservoirs as assessed in 1968. The dams did, however, enable intestinal schistosomiasis transmission to occur in the upper Awash Valley by changing its hydrology and introducing irrigation.

In West Africa, both urinary and intestinal forms of schistosomiasis became highly endemic in the Office du Niger area. The irrigation scheme had a mean prevalence of urinary schistosomiasis of 64.4%, compared to the river communities' 19.9%, and a mean prevalence of intestinal schistosomiasis of 53.9%, compared to the river communities' 1.9%.

Effects of Small Dams on the Prevalence of Bilharzia in Africa

Throughout the semi-arid areas of Africa, people have constructed many small earthen dams to provide irrigation water for dry season cultivation. While reports often blame these dams for spreading schistosomiasis, there is little evidence to substantiate such claims. Small dams in the semi-arid zone of West Africa get the greatest amount of attention. Although the dams built for dry-season irrigation extended the range of *Bulinus rohlfsi* snails, a common vector for urinary schistosomiasis in West Africa, this did not cause any noticeable increases in prevalence of urinary schistosomiasis.

Several studies in the same Sudano-Sahel ecological zone as northern Nigeria noted that evidence linking small earthen dams to schistosomiasis was lacking. In Burkina Faso, natural seasonal ponds infested by *Bulinus truncatus snails* are more common and dangerous locations of infections than are artificial reservoirs created by small dams.

In Cameroon, a nation-wide survey failed to find an association between the rate of schistosomiasis and small dams. Instead, temporary ponds and snail hosts adapted to low seasonal rainfall permits intense transmission of urinary and intestinal schistosomiasis throughout northern Cameroon, regardless of dams. Contrary to the experience in Nigeria, Burkina Faso and Cameroon, one study in northern Ghana showed small dams to be linked to schistosomiasis. Data collected during 1960-61 showed much higher prevalence of schistosomiasis in the eastern, more densely settled part of what was the Upper region of Ghana. The mean prevalence of urinary schistosomiasis in the region's eastern part was 19.8% in 15 districts without dams, 42.3% in 16 districts with dams 12 years old, and 52.0% for 6 districts with dams 3 years old. Areas in the western part of the region with few or no dams had infection rates under 10% in 6 districts, and 10 to 29% in 10 others. In contrast, prevalence was over 70% in 2 of the 3 western districts containing dams.

The Control of the Water-related Diseases

The control of the water-related diseases can be effected in a number of ways, some of which are mutually reinforcing. Three types of measures are distinguished:

- measures aimed at the pathogens: immunization, prophylactic or curative drugs;
- measures aimed at reducing vector densities or vector lifespan: chemical, biological and environmental controls;
- measures to reduce human/vector or human/pathogen contact: health education, personal protection measures and mosquito proofing of houses.

Of the above, environmental control measures are considered to be long-lasting and environmentally-sound. These include preventing or removing aquatic vegetation, lining canals with cement or plastic, regularly fluctuating water levels, periodic rapid drying of irrigation canals, preventing contamination of water bodies with faeces, supply of safe and clean drinking water, appropriate siting of housing of the farmers etc. For example, in Zimbabwe, in a communal small-holder irrigation project at Mushandike, adoption of a these measures resulted within three years in a drop of the infection rate from an initial 70 to 80% to virtually nil.

Potential Environmental Impacts of Darns and Reservoirs

The benefits of a dam project are flood control and the provision of a more reliable and higher quality water supply for irrigation, domestic and industrial use. Intensification of agriculture locally through irrigation can reduce pressure on uncleared forest lands, intact wildlife habitat and marginal agricultural land. In addition, dams create reservoir fishery and the possibilities for agricultural production on the reservoir drawdown area, which more than compensate for losses in these sectors due to the dam construction.

However, large dam projects cause irreversible environmental changes over a wide geographic area and thus have the potential for significant impacts. Criticism of such projects has grown in the last decade. Severe critics claim that because benefits from dams are outweighed by their social, environmental and economic costs, the construction of large dams is unjustifiable. In some cases, environmental and social costs can be avoided or reduced to an acceptable level by carefully assessing potential problems and implementing cost-effective corrective measures.

Damming the river and creating a lake-like environment profoundly changes the hydrology and limnology of the river system. Dramatic changes occur in the timing of flow, quality, quantity and use of water, aquatic biota, and sedimentation in the river basin. The area of influence of a dam project extends from the upper limits of the catchment of the reservoir to as far downstream as the estuary, coast and offshore zone. While there are direct environmental impacts associated with the construction of the dam (for example dust, erosion, borrow and disposal problems), the greatest impacts result from the impoundment of water, flooding of land to form the reservoir and alteration of water flow downstream. These effects have direct impacts on soils, vegetation, wildlife and wildlands, fisheries, climate and especially the human populations in the area.

Increased pressure on upland areas above the dam is a common phenomenon caused by the resettlement of people from the inundated areas and by the uncontrolled influx of newcomers into the basin catchment. On-site environmental deterioration as well as a decrease in water quality and increase in sedimentation rates in the reservoir result from clearing of forest land for agriculture, grazing pressures, use of agricultural chemicals, and tree cutting for timber or fuelwood.

The Impact of Dams on Flooding

The function of dams and reservoirs in flood control is to reduce the peak flows entering a flood prone area. Rather than maintaining high water levels for increased head or sustained water supply for irrigation, flood control operation requires that water levels be kept drawn down deliberately prior to and during the flood season in order to maintain the capacity to store any incoming floodwater. However, flood plains may be productive environments because flooding makes them so. Flooding recharges soil moisture and replenishes the rich alluvial soils with flood deposits of silt. In arid areas flooding may be the only source of natural irrigation and soil enrichment. Reduction or elimination of flooding has the potential for impoverishing flood recession cropping, groundwater recharge, natural vegetation, wildlife and livestock population in the flood plain which are adapted to the natural flood cycles. To maintain the productivity level of the natural systems, compensatory measures have to be taken, such as fertilization or irrigation of agricultural lands. In addition, when channelization measures reduce the frequency of flooding, the sediments entering the river systems from catchment areas upstream will be passed to the mouth of the river unless overflow areas are present downstream. Channel modification can result in a number of negative environmental impacts. Any measure that increases the velocity of flow increases the erosive capacity of the water. Although channel improvement can alleviate flooding problems in the treatment area, flood peaks are likely to increase downstream, thus simply transferring the problem elsewhere. Dikes built on the flood plain to exclude water from certain areas affect the hydrology of the area, and can have impacts on wildlife and livestock habitat and movement.

The Impact of Dams on Fisheries and Wildlife

Fishery alongside the rivers usually declines due to changes in river flow, deterioration of water quality, water temperature changes, loss of spawning grounds and barriers to fish migration. A reservoir fishery, sometimes snore productive than the previous fishery alongside the river, however, is created.

In rivers with biologically productive estuaries, both marine and estuarine fish and shellfish suffer from changes in water flow and quality. Changes in freshwater flows and thus the salinity balance in an estuary will alter species distribution and breeding patterns of fish. Changes in nutrient levels and a decrease in the quality of the river water can also have profound impacts on the productivity of an estuary. These changes can also have major effects on marine species which feed or spend part of their life cycle in the estuary, or are influenced by water quality changes in the coastal areas. The greatest impact on wildlife will come from loss of habitat resulting from reservoir filling and land use changes in the catchment area. Migratory patterns of wildlife may be disrupted by the reservoir and associated developments. Aquatic fauna, including waterfowl, amphibians and reptiles can increase because of the reservoir.

Socio-economic Impacts Irrigation Schemes

The objective of irrigation projects is to increase agricultural production and consequently to improve the economic and social well-being of the rural population. However, changing land use patterns may have other impacts on social and economic structure of the project area. Small plots, communal land use rights, and conflicting traditional and legal land rights all create difficulties when land is converted to irrigated agriculture. Land tenure/ownership patterns are almost certain to be disrupted by major rehabilitation works as well as a new irrigation project. Similar problems arise as a result of changes to rights to water. Increased inequity in opportunity often results from changing land use or water use patterns. For example, owners benefit in a greater proportion than tenants or those with communal rights to land. Access improvements and changes to the infrastructure are likely to require some field layout changes and a loss of some cultivated land.

Irrigation projects tend to encourage population densities to increase, either because of the increased production of the area or because they are part of a resettlement project. Impacts resulting from changes to the demographic/ethnic composition may be important and have to be considered at the project planning stage through, for example, sufficient infrastructure provision.

The most significant issue arising from large dam construction is resettlement of people displaced by the flooding of land and homes. This can be particularly disruptive to communities and insensitive project development would cause unnecessary problems by lack of inadequate compensation of the affected population. Human migration and displacement are commensurate with a breakdown in community infrastructure which results in a degree of social unrest and may contribute to malnutrition. As an example of the number of people displaced by the construction of a dam, filling of the reservoir behind the High Aswan Dam displaced 50000 to 60000 people in Egypt and some 53000 people in the Sudanese portion.

Changing land patterns and work loads resulting from the introduction or formalizing of irrigation are likely to affect men and women, ethnic groups and social classes unequally. Groups that use common land to make their living or fulfil their household duties, for example for charcoal making, hunting, grazing, collecting fuel wood, growing vegetables, etc. may be disadvantaged if that same land is taken over for irrigated agriculture or for building irrigation infrastructure. Women, migrants groups and poorer social classes have often lost access to resources and gained increased work loads. Conversely, the increased income and improved nutrition from irrigated agriculture may benefit women and children in particular. The most common socio-economic problems reducing the income generating capacity of irrigation schemes are:

- The social organization of irrigation operation and maintenance (O&M). Poor O&M contributes significantly to long-term salinity and waterlogging problems and needs to be adequately planned at the design stage to sustain the long-term development of the schemes.
- Reduced farming flexibility. Irrigation may only be viable with high-value crops, thus reducing extensive activities such as grazing animals, operating woodlots, etc.
- Changing labour patterns that make labour-intensive irrigation unattractive.
- Insufficient external supports such as markets, agrochemical inputs, extension and credit facilities.

User participation at the planning and design stages of both new schemes and the rehabilitation of existing schemes, as well as the provision of extension, marketing and credit services, can minimize negative impacts and maximize positive ones.

Alternatives to Mitigate the Negative Impacts of Irrigation Projects

Alternatives exist to mitigate adverse effects of irrigation development. Some of them are listed below:

- locating the irrigation project on the site where negative impacts are minimized;
- improving the efficiency of existing projects and restoring degraded croplands to use rather than establishing a new irrigation project;
- developing small-scale, individually-owned irrigation systems as an alternative to large-scale, publicly-owned and managed schemes;
- using sprinkler irrigation and micro-irrigation systems to decrease the risk of waterlogging, erosion and inefficient water use;
- using treated wastewater, where appropriate, to make more water available to other users;
- maintaining flood flows downstream of the dams to ensure that an adequate area is flooded each year, among other reasons, for fishery activities.

The Role of Wetland and the Impacts of Water Development Projects

Wetlands are wildlands of particular importance both economically and environmentally. The most important roles which wetlands perform are:

- Preservation of biological diversity: for many species of shrimp, fish and waterfowl, tidal and fresh water marshes, coastal lagoons and estuaries are of vital importance as breeding grounds as well as staging areas in their migration routes. All types of wetlands may harbour unique plants and animals.
- Production of goods: wetlands are among the most productive ecosystems in the world. Estuaries and tidal wetlands, in particular mangroves, are important nursery areas for most species of fish and shrimp which are later caught offshore. Shallow water areas are, in general, rich fishing grounds. Flood plains are important grazing areas for cattle and wildlife and vital spawning grounds for many fish species. Swamp forest may yield valuable timber.
- Production of services: wetlands can contribute to local rainfall and can be an efficient, low cost water purification system (herbaceous swamps), a recreation area (hunting, fishing, boating), a buffer against floods, and provide protection against coastal erosion by storms (mangroves).

Despite their importance, wetlands are under threat, in particular, from direct conversion of wetlands for agriculture and projects which affect the hydrology of a wetland, such as construction of dams, flood control, lowering of the aquifer drainage, and irrigation and other water supply systems.

The sections below describe some important wetlands in Africa, but the list is far from exhaustive.

Wetlands in the West African Sahel

The Senegal river waters the west of Mali, the north of Guinea, the north of Senegal and the south of Mauritania, the main Niger river stretches through Mali, the Republic of Niger and Nigeria. The frontiers of Niger, Nigeria, Cameroon and Chad meet in the Lake Chad. For all of these countries, river valleys and lakes are of the utmost economic importance because of their high productivity. Mauritania, Mali, Niger and Chad, whose territories are for the most part desert, draw their agricultural resources and the greater part of their animal protein from fishing and stock-raising in the river valleys. The states along the water courses are acting in concert to increase the use of water resources and avoid the hazards of drought. It is a question of ensuring mastery over the flood waters to grow rice and other cereals, and of using surface water for irrigation. These operations inevitably tend to modify ecological conditions. In practice this will lead to significant losses of aquatic habitats and to a noticeable decrease in fish and waterfowl.

Regions such as the southern part of the Sahel, with a strongly seasonal rainfall regime and yet with sufficient rainfall to support seasonal agriculture and pastoralism, support large numbers of people. Flood plains, swamps and lakes provide a range of ecological resources and economic opportunities. Without wetlands, the drylands of the West African Sahel would be both less productive and more hazardous as a place for people to live.

In the semi-arid zone of Western Africa, patterns of rainfall and river flow are strongly seasonal. In northern Nigeria, most of the rainfall occurs in just three or four months, between June and September. During this short rainy season, precipitation exceeds evapotranspiration and runoff occurs. Savannah rivers run strongly, but start to shrink as the rains end. During the wet season rainfed agriculture is possible, and there are extensive grasslands providing relatively nutritious grazing for livestock away from the river valleys. Once the rains end, these resources also dry up, and pastoralists concentrate on the remaining wetlands in river valleys or larger wetlands such as the delta of the Senegal River the Niger Inner Delta in Mali or Lake Chad.

It is also only in these areas that agriculture can continue into the dry season. In many areas rice is planted, either in the rains before the floods arrive, or as the floods recede. Flood-recession crops such as sorghum or beans are planted as the waters recede, and farmers dig wells or use remaining pools to irrigate small gardens using buckets, shadoofs or, more recently, small pumps. At the same time, other economic activities such as fishing are also linked to the changing flood. Many fishes move laterally out of the riverbed pools into the flood plain to breed, and their offspring is caught as the water retreats. Later in the dry season, the residual pools in the riverbed are themselves fished. The high fisheries productivity of most of the seasonally inundated flood plains is fostered, at least in part, by the nutrient-rich dung left by the grazing animals during the previous dry season.

In valleys such as the Senegal or the vast flooded plains of the Niger Inner Delta, the annual cycle of farmers, herders and fishers is closely linked to the seasonal cycle of flooding. The flood plain of the Senegal stretches up to 30 km in width, and runs 600 km downstream of Bakel. It covers a total of about 1 million ha and supports farmers, pastoralists and fishing communities. Up to half a million people depend on the flood-related cropping in the *'waalo'* land of the flood plain.

There is growing evidence that large-scale capital intensive water development schemes do provide neither the range of foodstuffs nor the economic return of traditional systems. Studies, comparing the efficiency per unit of water of traditional extensive systems of cultivation, grazing and fishing in the Niger Inner Delta with the intensive modern project of the Office du Niger, showed that both systems produce about the same gross profit margin, even when the running costs and management charges for the irrigation scheme are taken into account. However, the extensive system produces meat, milk, fish and rice compared to the rice-only irrigated system. More importantly, when the interest charges arising from the irrigation scheme are taken into account, the net profit from the irrigated rice turns into a loss of \$0.65/100 m^3 of water, whilst the extensive traditional methods benefit from the 'free services of nature' and turn in a net profit of \$0.42/100 m^3 at 1984/85 prices.

The Hadejia-Nguru Wetlands

The Hadejia-Nguru Wetlands concern a part of the flood plain of the Komadougou-Yobe river basin in the Lake Chad basin in the north-east of Nigeria and are home to probably about a million people. The wetlands have formed where the waters of the Hadejia and Jama'are rivers meet the lines of ancient sand dunes aligned northeast-southwest. An area of confused drainage has formed here, with multiple river channels and a complex pattern of permanently and seasonally flooded land and dryland. The wetlands are nationally and internationally important for migratory waterfowl. The wetlands support extensive wet-season rice farming, flood-recession agriculture and dry-season irrigation. The flood plain also supports large numbers of fishing people, most of whom also farm, and is grazed by very substantial numbers of Fulani livestock, particularly cattle, which are brought in from both north and south in the dry season. There is also an important dispatch from the wetlands of fuelwood and fodder for horses. In the past, much of the rice, as well as fish and birds, was traded out of the area. This has changed, but there is now a strong export of other agricultural products, for example peppers, wheat and fuelwood. The economic value of production from the wetlands is very large, many times greater than that of all the irrigation schemes for which the inflowing rivers are dammed, diverted and their waters used. There are natural changes, for example the impacts of drought, that have serious implications for the future of the wetlands and the sustainability of their production systems. There are also major economic changes within the wetlands themselves. The extent of irrigation has greatly increased over the 1980s, largely as a result of the advent of small petrol-powered pumps and the ban on the importation of wheat in 1988. As the use of small pumps spreads, conflicts are beginning to emerge between farmers and pastoralists, and between small and large farmers for access to land.

The wetlands have also been affected by developments elsewhere in the river basin. The construction of the Tiga Dam on a tributary of the Hadejia river in the early years of the 1970s has exacerbated the effects of the low rainfall of the last two decades. The result has been a reduction in the extent of flooding in the wetland. Construction of a dam on the Hadejia river just above Hadejia town to provide short-term storage of water to irrigate the Hadejia Valley Project Phase 1 began, in the early 1980s, but was stopped for several years because of financial problems. The main dam was completed in 1992, soon after work restarted on the project. The dam has created a large shallow lake upstream and it will probably have a major effect on the timing and extent of flooding in the wetlands.

Most of the dams, irrigation schemes and water resources plans for the Yobe basin were prepared in the 1970s and early 1980s, using data for the relatively wet period up to 1973. The post-1972 drought has reduced the proportion of rainfall which runs off to the rivers. The 1988 flood at Hadejia was probably one of the largest for some years and it was augmented by the failure of the dam at Bagauda.

The Hadejia-Nguru wetlands have long been known as a centre of fish production. Upstream hydrological developments induced by irrigation projects threaten to degrade this important resource. Studies of flood plain fisheries have shown that fish production is closely related to flood extent. The existing and planned dams upstream of the Hadejia-Nguru wetlands are likely to have a serious impact on fisheries. Despite the lack of information specific to the Hadejia-Nguru wetlands, there are enough studies from other flood plains affected by hydraulic works to show that the effects of dams on fish communities are likely to be serious. The dams are likely to bring changes in river flow, loss of habitat, blocking of channels, changes in silt loading, plankton abundance and temperature which are likely to affect fish communities.

The economic value of fish production from the flood plains adds weight to the argument in favour of maintaining the annual flooding of the wetlands. Moreover, the significance of fishing goes beyond its value in monetary terms. Fishing plays an important role in the flexibility and adaptability of the rural economy in the flood plains. A reduction in this flexibility through degradation of the fishery resource may have serious repercussions on the ability of communities to adapt to fluctuations in their environment. Many people are involved in the fisheries and so the social consequences of any appreciable reduction in productivity will be felt throughout the area. Degradation of the fisheries may also affect other sectors of the rural economy. Most people who fish also pursue other activities-such as farming, livestock rearing, manufacturing of crafts or trading-and the loss of, or reduction in one component of the household economy is likely to affect activities in other sectors. There will also be 'downline' effects on fish processors, fish dealers, customers and consumers.

In addition to producing fuelwood, the forest reserves and bushland of the flood plains yield important non-timber forest products that are significant to the livelihoods and subsistence of local communities. Some, including leaves, are important marketed commodities that generate substantial income. *Doum* palm leaves are either processed into mats and other products or sold as raw material. The harvesting and processing of doum palm leaves is a dry season activity, and many people migrate to the wetlands to harvest the palm. Mat-making from doum is also a specialized activity of many flood plain villages. Mats and other doum products, for example rope and baskets, are sold locally or exported to other regions. *Baobab* leaves are used widely as an ingredient for soups and stews and are especially important as a 'drought food'. Honey, produced by local beekeepers, is a highly valued commodity.

Since 1985, the area has been the focus of the Hadejia-Nguru Wetlands Conservation Project. This project has been run jointly by the Nigerian Conservation Foundation, IUCN (International Union for the Conservation of Nature), the Royal Society for the Protection of Birds and the International Council for Bird Preservation (now renamed Birdlife International). In 1990 a major development project was started by the European Community that included the eastern part of the area. The North East Arid Zone Development Project (NEAZDP) has a very substantial budget to generate village-based development initiatives. Attention has tended to be directed in particular to the potential resources of the wetlands.

Wise use of the wetlands of the Hadejia-Nguru wetlands demands a proper understanding of the environmental and socio-economic changes that are occurring and of those that may be predicted. Understanding of the impacts of changes inside and outside the flood plain is far from easy, and prediction of future impacts is even harder. However, without such understanding and prediction, effective planning and management is impossible.

The economic importance of the flood plains suggests that benefits it provides cannot be excluded as an opportunity cost of any scheme that diverts water away from the flood plain system. Policy makers should be aware of this problem when designing water development projects in the river system. Further analysis is also required of the type of 'regulated flood projects' regime, which could maintain much of the flood plain system intact while still allowing some upstream water developments. Further investigation of all the economic benefits provided by the wetlands is also needed, and the sustainability of production within a flood plain area should be more thoroughly examined.

Effects of the Jonglei Canal on the Sudd Swamps

In the southern of Sudan, the Nile discharges its water into the great wetlands of the Sudd, a network of channels, lakes and swamps in which as much as half of the inflowing water is disappears through evaporation. The Jonglei Canal was designed to bypass the Sudd and direct downstream a proportion of the water that is 'lost' from the Nile each year by spill and evaporation in the swamps. The projected dimensions of the canal are as follows: a width from bank to bank of about 75 metres, a channel bed-width averaging 38 metres, a depth varying from 4 to 8 meters, and a length of 360 km, over twice the length of the Suez canal. Jonglei is a small Dinka village close to the Atem channel at a point where the canal alignment was planned to begin. Although the offtake will now be further south at Bor, the canal is still so named and Jonglei has given its name to a province as well.

The canal has not been completed, but detailed surveys were undertaken to determine a whole range of effects, many of which will be shown to be disadvantageous to the inhabitants of the Jonglei Area. Some of the effects are described below.

The river-flooded grasslands are an essential seasonal resource during the driest months of the year. Not only is there drinking water available in the rivers, but the process of seasonal inundation itself produces species of grasses which sustain the herds from about January until April. There are no other alternatives as the grasses of the high land are exhausted or reserved for the livestock (mainly smaller stock), held by the few people who elect to remain behind, and the rain-flooded grasslands have become woody and unpalatable and produce little or no regrowth after burning. It follows that the river-flooded grasslands are crucial to the pastoral economy at this time of the year. It is, however, just these grasslands that may be reduced by the operation of the canal.

The water benefit of the canal downstream will be around 4 km^3/year and according to some estimates even an extra water flow of up to 10 km^3/year may be reached. These quantities are a substantial percentage of the average 'losses' by the evapotranspiration, the natural production of river-flooded grasses being a function of the annual fluctuation in river discharge and thus of the annual variation in area flooded. In other words, to the local inhabitants these are not losses in water at all, though the waters are excessive and the cause of damaging floods, as in 1964.

The floods of the 1960s, reaching a peak in 1964, caused great damage to human interests. On the Zeraf island alone it was reckoned that 130000 cattle were lost owing to exposure and lack of grazing since practically the whole area remained under water for a long period. Similar disastrous effects occurred west of the Bahr el Jebel in the vicinity of Adok. It follows that any reduction in peak flows could be protective and beneficial. The same model can be applied to give some indication of the effect of the canal on areas of flooding. The figure of 25 million m^3/day for a canal diversion may reduce the area of flooding by about 19% at a 1964 peak discharge.

The established fisheries of some large lakes in the Sudd are said to have been adversely affected by increased water depth. but, overall, the flooding of the 1960s has multiplied the number of perennial lakes in the system and, thereby, the fishing potential. A severe decrease in the discharge into the Sudd resulting from the Jonglei canal would bring about the total disappearance of many lakes in the papyrus zone and reduce others to the status of seasonal lagoons, with a serious loss of year-round fish and fishing potential. There is, however, likely to be considerable disadvantage to the people of the Zeraf Island and those living west of the Bahr el Jebel, in that mainstream traffic will follow the canal and the old western landing places will be ill-served. In the past, moreover, river traffic has been a major factor in keeping the channels open. Oil prospecting is likely to restart once peace has been restored and this may mean that the companies concerned will wish to keep channels clear. However, if discharges drop to the low figures prior to 1961, the canal could become too shallow for commercial traffic and for the movement of fisheries barges.

The canal will in many areas drive a barrier between wet season villages and dry season grazing grounds along the river channels and therefore dislocate the pastoral cycle. Many people living east of the canal will have to cross it with their livestock when regrowth from rain-flooded grasslands is exhausted and they have to move westwards to the river-flooded grasslands of the Nile. Reinforced structures at various points along the canal are needed to facilitate the crossing of livestock without damage to the embankments and to provide suitably designed boats more efficient than the usual 'dug-out' canoe. Crossing the canal will present a massive logistical problem and besides, raises questions of land ownership among those who may need to cross the canal and cross each others' territory in order to do so. There exists a kind of 'Jonglei Controversy'. The criticism of the environmentalists are many but can be segregated into charges that the Jonglei Canal will drastically affect climate, groundwater recharges, silt and water quality, the destruction of fish and changes in the lifestyle of the Nilotic people. However, other studies claim that the positive effects will counterbalance by far the negative effects. As is the case with the Hadejia-Nguru wetlands understanding and prediction of the impacts is very difficult. However, without such understanding and prediction, effective planning and management is impossible.

Regional Aspects of Environmental Impacts and 'Hot Spots'

This section summarizes the regional outlook for the main African sub-regions with regard to the impact of irrigation on the environment. For each of the sub-regions, environmentally salient features, particularly in relation to irrigation development issues, are presented. Wherever possible, environmental 'hot spots' are identified and described.

The Arid North African Sub-region

The North African sub-region lies in arid or semi-arid zones where the water resources are minimal and where evaporation and seepage losses are very large. The sub-region includes two main zones: the Nile Basin in the east and the western part (Morocco, Algeria and Tunisia). This ecological zone is fragile and agricultural production is regulated by alternating periods of water surpluses and deficiencies. Irrigation is the main alternative to cover the food requirements of the increasing population per unit of agricultural land.

The 'hot spots' of the sub-region are the main rivers located in the arid zones threatened by the irrigation-induced salinization of the soils and more generally the degradation of irrigated lands resulting from poor irrigation management and practices. Field drainage and removal of drained water from the irrigated zone is necessary to limit the risk of soil degradation and salinization. Establishing field drainage is costly, as is the provision of a main drainage network. Moreover, disposal of drainage water represents a major problem. The concentration of salt increases gradually from upstream to downstream as a result of the drainage water inflows.

In the Mediterranean coastal zone, reduction of flow systematically induces sea water intrusion problems.

The Sudano-sahelian Belt

The natural capital of this ecological zone is the most fragile, evincing most of the negative effects of irrigation projects on the environment due to the poor soils, extremely variable rainfall and high risk of drought. Soil degradation has substantially increased the risk of decertification because of mutually reinforcing factors including: loss of organic matter and nutrients, soil structure deterioration and surface crusting, which in turn decreases water infiltration and retention, aggravated by the irrigation-induced salinization.

The environmental degradation has been both a cause and a consequence of poverty, with the Sudano-Sahelian belt comprising some of the poorest countries of the world.

Sudano-Sahelian societies face a formidable challenge which makes the whole belt an environmental 'hot spot'. Pressure is likely to be high on the river valleys and major wetlands such as the Niger Inner Delta in Mali, the wetland and flood plains in the Lake Chad basin (Hadejia-Nguru, Yaere and the Sudd swamps in the Nile basin in southern Sudan.

Humid West Africa

The natural capital of this ecological zone is relatively favourable in terms of climatic conditions: high and regular rainfall, soils of reasonable quality. However, the high population growth during the last decades has placed the environment under serious stress. About half of the total land area is cultivated under reasonably good conditions with a much lower climatic risk than in the Sahel. Forest land has shrunk to less than a third of the total area, and what remains is decreasing at an alarming rate of 1%, the fastest rate in tropical

Africa. The environment of the fragile coastal ecosystems is also threatened by industrial and urban development with increasing pollution levels particularly in the Niger delta of Nigeria. A major part of the biodiversity capital of the sub-region is at risk.

Any upstream irrigation project requires special care in order to avoid negative impacts on the wetlands, mangroves and lagoons located in the coastal zone in the Guinea Gulf, from Guinea Bissau to the Niger delta in Nigeria. This zone is likely to become a continuous urban megalopolis with a population of over 50 million people on 500 km of coastal line. The current development of private small-scale irrigation projects, using groundwater for horticultural crops, could contribute to increase the intrusion of saline waters due to over exploitation of coastal aquifers.

The Congo/Zaire Basin

About three-quarters of the total surface area of the sub-region would theoretically be cultivable but a major part of it is under tropical rainforest. Overall pressure to clear the rainforest is still relatively low, except at the periphery of the sub-region where it interfaces with areas of high population density.

Land currently cultivated represents about 15% of the total area. Agricultural activities are relatively less important in the sub-region compared to the rest of the continent. They are focused on supplying a growing urban market and on permanent plantations.

Irrigated areas are marginal compared to the huge potential of land and surface waters. The irrigation development will have a minimal impact on the environment. Global environmental problems faced by the Congo/Zaire basin are less severe than those of the other sub-regions, although its future development will present a serious challenge. In particular, countries need to preserve the primary rainforest for global biodiversity and climatic reasons.

East Africa

The good soils in the eastern African highlands have favoured the development of intensive agriculture, although soils require conservation measures because of steep slopes. Less favourable lands are cultivated under arid and semi-arid conditions. Forests cover less than 20% of the total area of the sub-region. Due to land scarcity, the primary rainforests with their unique biodiversity are at risk.

Due to the pressure of population on arable lands, the environment is at risk particularly in Kenya with a ratio of 0.2 ha per person. To cover the food deficits, areas under irrigation are increasing. Permanent intensive cropping is the current pattern in favourable highlands, but degradation is high under low-input technology and without adequate erosion control measures. In Tanzania, the central area and the Lake Victoria region represent areas of high population density and areas with a reported high degree of land degradation.

Ethiopia has a very large water resources potential. The development of this resource has been impeded for decades, first by agreements made by colonial powers and shell by political instability. The Ethiopian Blue Nile and other tributaries contribute over 80 % of the water in Sudan and Egypt. The mobilization of this potential would have to take into account environmental and basin issues to mitigate the impact on downstream users.

Southern Africa

The natural capital of the sub-region is very rich in terms of biodiversity and production potential, although large areas are under semi-arid and arid conditions with moderate to high risk of drought.

In some countries, particularly South Africa, past policies have had a negative impact on the environment by encouraging agricultural development through high subsidies on farm inputs and irrigation development without stimulating enough soil and water conservation.

Almost half of the total areas of the sub-region is cultivated with reasonably good soils but climatic conditions are highly variable with a risk of recurrent droughts. To mitigate this risk and to cover the food deficit, areas under irrigation are increasing without significant impact on the environment.

10

Sustainable Water Resource Management

The current peace process offers a special opportunity for all the nations in the Middle East to abandon the existing status of belligerency, confrontation, non-cooperation, and polarization. The ultimate objective is to arrive at a comprehensive, just and lasting peace in the whole region under which all the peoples of the area can together develop the area and promote progress and prosperity. Water is a major issue that can catalyze the peace process or inhibit it. After more than five years of meetings and negotiations, the gap in the positions among regional parities is still as wide as ever. The region's hydrologists and politicians are still talking on different wavelengths. This chapter will focus on the Israeli-Palestinian water disputes in the groundwater aquifers and Jordan River. We realize that water is a particularly sensitive and critical issue for all parties to the conflicts. But we also believe that finding a common understanding of water issues in the Middle East would go far to enhance the possibilities of achieving stability in the region. Conversely, failure to reach these common grounds will, most definitely, obstruct any efforts to attain this goal. There is no alternative to an honest and forthright discussion of the water issues and to exposing the current unsustainable reality of mismanagement, inequities, and the outright denial of the Palestinian's inalienable right to their resources.

Water Resources

Water does not recognize political boundaries and, as such, it is difficult to delineate Israeli and Palestinian surface and ground water resources. Nevertheless, we outline here the water resources in the West Bank and Gaza Strip, as well as those in Israel.

Surface Water

Surface water is that which flows permanently in the form of rivers and wadis or that which is held in seasonal reservoirs. The Jordan River is the only permanent river which can be used as a source of surface water in Palestine. The Jordan River is 360 kms long with a surface catchment area of which 18,300 [km^2] lie upstream of the Lake Tiberias outlet. The average annual flow of this river is about 1311 MCM (Haddad, 1997). The Jordan River initiates from three main springs: The Hasbani in Lebanon, the Dan in occupied Palestine, and the Banias in the Syrian Golan Heights to form the Upper Jordan river basin. The water of this basin flows southward through Lake Hula towards Lake Tiberias. In the absence of irrigation extraction, the Jordan River system would be capable of delivering an average annual flow of 1,850 MCM to the Dead Sea. The riparians of the Jordan River are Lebanon, Syria, Palestine, and Jordan. Only three percent of the Jordan River's basin fall within Israel's pre-1967 boundaries. Average precipitation for Upper Jordan and Lake Tiberias averages 1,600 mm and 800 mm respectively. The lower basin, around the Dead Sea has a desert climate characterized by scarce rainfall. The Jordan River is progressively more saline and less usable towards the Dead Sea. The Jordan River system satisfies about 50% of Israel's and Jordan's water demand; Lebanon and Syria are minor users, meeting 5% of their re-combined demands via the Jordan.

Downstream of Tiberias is the Lower Jordan river basin, which joins the Yarmouk and the Zerka Rivers originating from Syria and Jordan in the east. The outlet of this basin is toward the Dead Sea in the south. As a result of water diversion from the upper Jordan by the Israel, there is no fresh water to flow downstream of Tiberias. In normal years Israel allows a flow downstream from Lake Tiberias of just 60 MCM of water basically consisting of saline springs which previously used to feed the lake, and sewage water. These are then joined by what is left of the Yarmouk, by some irrigation return flows, and by winter runoff, adding up to a total of 200-300 MCM. Both in quantity and quality this water is unsuitable for irrigation and does not sufficiently supply natural systems.

Flood Water Flow

Surface flood runoff in the West Bank is mostly intermittent and probably occurs when the rainfall exceeds 50 mm in one day or 70 mm on two consecutive days. The runoff is estimated at about 64 MCM/yr in the West Bank. Streams flowing from the west towards the Jordan Valley recharge shallow aquifers such as Wadi al-Qilt, Auja and Wadi al-Far'a (Assaf, 1991). The flood wadis can be divided according to the flood flow direction as follows:

1. The eastern and northeastern flood wadis that have an average total annual flood flow volume of about 18.57 MCM/yr.
2. The western flood wadis that have an average total annual flood flow volume of about 17.91 MCM/yr.

In addition, there are small-scale wadis that discharge a total flood water volume that may reach 15 MCM/yr during the very wet seasons. In the Gaza Strip, runoff water is collected in small wadis and valleys within the area. Wadi Gaza is the most important. It drains 3,500 [km.sup.2] of the northern Negev. The northeastern part of the Gaza Strip, with loessial and alluvial soils, also contains some wadis. These soils have a low infiltration capacity; therefore, there are many surface run-offs during intensive rainfall.

West Bank Aquifer Systems

Groundwater is the major source of fresh water supply in the West Bank and Gaza Strip. In the West Bank, the aquifer system is comprised of several rock formations from the Lower Cretaceous to the Holocene geologic age. Most of the formations are comprised of carbonate rocks (mainly limestone, dolomite, chalk, marl, and clay). The aquifer system is recharged from rainfall in the West Bank. The main recharge areas are along the upper mountain slopes and ridges. The annual rainfall in the West Bank is estimated at 3,000 MCM (Abu Mayleh, 1994). Around 600-650 MCM of this rainfall is estimated to infiltrate the soil to replenish the aquifers annually. There are two general directions for the groundwater of the West Bank Aquifer system, east and west. The groundwater basins are recharged directly from rainfall on the outcropping geologic formations in the West Bank mountains (forming the phreatic portion), while the greatest part of the storage areas is located in the confined portions. The phreatic portions constitute the subsurface area under the West Bank mountains where the Palestinians dug their groundwater wells to tap the shallow unconfined aquifers. The Israelis, however, dug their wells to tap the confined aquifers whose quality and quantity are better. The West Bank aquifer system is classified according to flow direction into:

1. The Western Aquifer System, which is the largest, has a safe yield of 360 MCM per year (of which 40 MCM brackish). Eighty

percent of the recharge area of this basin is located within the West Bank boundaries, whereas 80 percent of the storage area is located within Israeli borders. Groundwater flow is towards the coastal plain in the west, making this a shared basin between Israelis and Palestinians. The groundwater being mainly of good quality, this source is largely used for municipal supply. Israelis exploit the aquifers of this basin through 300 deep groundwater wells to the west of the Green Line, as well as through Mekorot (the Israeli water company) deep wells within the West Bank boundary. Palestinians, on the other hand, consume only about 7.5 percent of its safe yield. They extract their water from 138 groundwater wells tapping the Western Aquifer System (120 for irrigation and 18 for domestic use) in Qalqilya, Tulkarm, and West Nablus. There are 35 springs with an average flow discharge exceedin g 0.1 L/s located in this aquifer system.

2. The Northeastern Aquifer System has an annual safe yield of 140 MCM (of which 70 MCM brackish). Palestinians consume only about 18% of the safe yield of their aquifers in the Jenin district and East Nablus (Wadi Al Far'a, Wadi El Bathan, as well as Aqrabaniya and Nassariya) for both irrigation and domestic purposes. There are 86 Palestinian wells in this aquifer system (78 for irrigation and 8 for domestic use). The general groundwater flow is towards the Bisan natural springs in the north and northeast.
3. The Eastern Aquifer System has a safe yield of 100-150 MCM per year (of which 70 MCM brackish). It lies entirely within the West Bank territory and was used exclusively by Palestinian villagers and farmers until 1967. After 1967 Israel expanded its control over this aquifer and began to tap it, mainly to supply Israeli settlements implanted in the area. The most important springs in the West Bank are in this basin. Seventy-nine springs with an average discharge greater than 0.1 L/s provide 90 percent of the total annual spring discharge in the West Bank. There are 122 Palestinian groundwater wells in this aquifer system (109 for irrigation and 13 for domestic use).

Gaza Coastal Aquifer

The main Gaza Aquifer is a continuation of the shallow sandy/ sandstone coastal aquifer of Israel (shared aquifer) which is of the Pliocene-Pleistocene geological age. About 2200 wells tap this aquifer with depths mostly ranging between 25 and 30 meters. Its annual safe yield is 55 MCM (GTZ, 1998), but the aquifer has been over-pumped at the rate of 110 MCM resulting in a lowering of the groundwater, sea level and saline water intrusion in many areas. The main sources of salinity are deep saline water intrusion from deeper saline strata, sea water intrusion, and return flows from very intensive irrigation activities. Another water resource within the area is the Sea of Galilee; much of its water and its surface water is delivered directly to the population of settlements in the vicinity via the National Water Carrier. Some additional relatively smaller aquifers are found in the Western Galilee, in the Golan, in the northern valleys, in the Jordan Basins and in the Arava desert.

Water Quality

Palestinian's share in the River's water cannot be used because they have no access to the Jordan River due to military closure by the Israelis since 1967. Different riparians took their needs from the Jordan River basin and the small quantity that can reach the Palestinian riparian in the West Bank is of poor quality. The salinity of the lower Jordan River reaches up to 5,000 parts per million (ppm). This deterioration is due to the over-exploitation of the shallow aquifer tapped by wells, scarcity of rainfall and irrigation return flow. The deep Israeli wells in the Jordan Valley have good water quality since they tap the Lower Cenomanian aquifer system. The Israelis are currently consuming 40 MCM/yr. (Oslo II, 1995) of water from their deep groundwater wells in the Jordan Valley.

The water quality from the West Bank aquifers is quite good except in certain areas in the Eastern aquifer. However, most of the water extracted from the Gaza aquifers is of poor quality due to high salinity (reaching 1500 ppm in some cases) and the high level of nitrates (reaching more than 350 mg/l) that falls below health standards set by the World Health Organization. However, there are a limited number of water lenses under Gaza, which are of fresh water quality. These lenses are situated around the Israeli settlements in the Gaza Strip and thus are not accessible to the Palestinians even after Autonomy. Over-pumping of the Gaza aquifers has resulted in seawater intrusion and high salinity levels. In southern Gaza, groundwater salinity has been rising by 20 mg/l/year, and the water table has been declining by about 0.2 m per year (PEPA and Euroconsult/Iwaco, 1995). Chemically contaminated water runoff seeping into aquifers has also resulted in high nitrate levels.

Roots of the Water Conflict

To devise a solution to the water conflict, it is extremely important to look at its roots, which go back to the end of the past century when the Zionist movement initiated its plans for creating a Jewish homeland. In 1875, it was proposed that such a homeland should encompass Palestine, the Negev and parts of Jordan, with the water resources so that it could absorb 15 million Jews. After the declaration of the British mandate in 1922, the Jewish Agency formed a special technical committee to conduct studies of the utilization of water and irrigation of unarable and desert land. Most of the studies conducted were used to evaluate water plans designed by both the Jewish Agency and the United Nations Partition Plan of Palestine. The Arabs found it imperative to protect their water resources and, thus, began to design their own plans. Rising political tension in the region and the lack of a solution acceptable to all parties exacerbated the situation, which eventually exploded into several rounds of wars.

Two important water-related events characterize the British mandate period from 1922 to 1948, namely the Rutenberg Concession and the lonides Plan. In 1926, the British High Commissioner granted the Jewish-owned Palestine Electricity Corporation, founded by Pinhas Rutenberg, a 70-year concession to utilize the water of the Jordan and Yarmouk Rivers to generate electricity. The concession denied Arab farmers the right to use the water of the Yarmouk and Jordan Rivers upstream of their junction for any reason whatsoever, unless permission was granted by the Palestine Electricity Corporation. In 1937, the government of Great Britain assigned M. lonides, a hydrologist, to serve as the Director of Development for the East Jordan Government. His actual task was to conduct a study of the water resources and irrigation potentials of the Jordan Valley Basin. This study served as the main reference in the preparation of the proposed United Nationals Partition Plan of Palestine Published in 1939, the lonides Plan made three recommendations. Firstly, the Yarmouk flood waters were to be stored in Lake Tiberias. Secondly, the stored waters in Lake Tiberias, plus a block quote quantity of 1.76 CM/s of the Yarmouk River water diverted through the East Ghor canal, were to be used to irrigate 300,000 dunums of land east of the Jordan River. Finally, the secured irrigation water of the Jordan River system, estimated at a potential of 742 MCM, was to be used primarily within the Jordan Valley Basin. The Jewish agency was not satisfied with the findings or recommendations of Ionides. Following the 1948 war, Israel launched a Seven Year Plan aimed at diverting the Jordan River water south toward the Negev desert. In September 1953, the construction of the National Water Carrier began. The diversion originated at the Banat Yacoub Bridge in the demilitarized zone between Israel and Syria. After Arab objections to the excavation process, a temporary freeze on the work was announced and the United Sates presented another plan in another attempt to solve the region's water dispute. The Johnston plan, which was prepared under the supervision of the Tennessee Valley Authority, included water distribution quotas for the Jordan Valley Basin, estimated at 1287 MCM annually, among the riparian states.

The period between October 1953 and July 1955 was a stage of negotiating and bargaining over the allocation of the Jordan River waters. By the end of 1955, the Johnston Plan had become more favourable to Israel, whose share rose to 450 MCM, while Jordan's share dropped to 720 MCM. The failure to reach a regional agreement reinforced each country's allocation to proceed independently. In 1958, Israel reinitiated the National Water Carrier project but with some technical changes; also, the Seven-year Plan was replaced by a ten-year Plan. Arab reaction to Israel's National Water Carrier was to build dams on tributaries of the Jordan and Yarmouk Rivers, thus reducing the water flow to Israel. In 1965, Syria began building dams to divert water from the Banias and Dan Rivers in the Golan Heights. Israel sent its fighter planes to destroy the work sites. No regional water plans were devised after the Johnston Plan of 1954, which allocated the water between the riparians based on the irrigable areas within the waters hed line.

A West Ghor canal was included in the plan to provide Palestinians with Jordan River water that translated into 250 MCM per year. This project was never implemented. Following the 1967 war, Israel secured control over the headwaters of the Jordan River. Before 1967, the Palestinians had 720 groundwater wells for agricultural and domestic purposes. Soon after the occupation, Israel imposed a number of military orders to control Palestinian water resources. On August 15, 1967, the Israeli military commander issued Order No. 92, in which water was considered as a strategic resource. This order was followed by numerous other orders aimed at making basic changes in the water laws and regulations in force in the West Bank. Under Military Order No. 158 of 1967, it was not permissible for any person to set up or to assemble or to possess or to operate a water installation unless s license had been obtained from the area commander. This order continues to apply to allow wells and irrigation installations. Th e area commander can refuse to grant any license without the need for justification. These orders were followed by numerous military orders — No. 291, No. 457 or 9172, 484 of 1972, 494 of 1972, 715 of 1977 and 1376 of 1991-to achieve complete control over Palestinian water resources. Immediately after the end of the war, Israel destroyed 140 Palestinian water pumps in the Jordan Valley and made it difficult to obtain permits for new wells. Despite the rapid increase in population and demand on water, Israel, since 1967, has granted Palestinians of the West Bank only five permits for new water wells. All were to be used exclusively for domestic purposes. New water wells for agricultural purposes in the West Bank were also restructured to three permits.

Water Consumption In Israel and Palestine

Israel has restricted Palestinian water usage and exploited Palestinian water resources. Presently, more than 85% of the Palestinian water from the West Bank aquifers is taken by Israel, accounting for 25.3% of Israel's water needs. Palestinians are also denied their right to utilize water resources from the Jordan and Yarmouk Rivers, to which both Israel and Palestine are riparians. At present, Israel is drawing an annual 685 MCM from the Jordan River.

As a result of Israeli policies, Palestinians are permitted to utilize 238 MCM of the water resources to supply 2,895,683 Palestinians in both the West Bank and Gaza strip with their domestic, industrial and agricultural needs. By comparison, 5,757,900 Israelis are utilizing 1959 MCM. On a per capita basis, water consumption by Palestinians is 82[m^3] compared to 340 [m^3] for Israelis. It should be added that Jewish settlers in both the West Bank and Gaza Strip consume huge amounts of the scarce Palestinian water resources. The 5,500 settlers in the Gaza Strip consume 10 MCM/yr for all purposes, whereas the one million Palestinians within Gaza consume approximately 113 MCM/year (Nassereddin, 1997). In the West Bank, Jewish settlers are consuming 57.3 MCM per year (PWA, 1997). In the West Bank, Jewish settlers are consuming 57.3 MCM per year (PWA, 1997), while Palestinians are struggling to connect the remaining 25 percent of the Palestinian population to household water-distribution systems. Jewish settlers in the West Bank and Gaza Strip receive continuous water supply, largely from groundwater wells in the Palestinian Territories.

In both Palestine and Israel, the agricultural sector uses about two to three times the municipal water consumption. While the agricultural sector in Palestine contributes between 15 and 20 percent of the GDP, it contributes only 1.8 percent to the GDP in Israel. Irrigated area in the Palestinian Territories covers approximately 201,358.00 dunums, of which 94,727.5 dunum are located in the West Bank, mainly in the Jordan Valley, Jenin, and Tulkarm areas. On the other hand, the irrigated area in Israel increased by 340,000 dunum from 1970-1990 (Al-Musa, 1997). Irrigated area in Palestine consumes 151 MCM of water, whereas in Israel it consumes 1,252 MCM. About 64.5 percent of the total irrigated areas in the West Bank are used for vegetables. The Israelis use irrigated areas for citrus, avocado, mango, grapes, apples, peaches, bananas, dates, wheat, corn, cotton, peanuts, potato, vegetables, flowers and flower bulbs that consume huge quantities of water, but the Israeli government supports the farmers to grow such crops.

There is a wide variation in water consumption for domestic purposes between Palestinians and Israelis, as the per capita water use is 30 [m^3] for domestic and industrial purposes for Palestinians in comparison to 100 [m^3] for Israelis for domestic purposes. During summer months, most Palestinian communities experience extended water shortages that last for weeks. For example, during the summer month of 1998, Israel supplied the Palestinian residents of Hebron district with 8500 [m^3] of water per day which is half the regular allotment of 1700 [m^3] of water promised to the city and nearby areas under the items of water agreements. The problem is exemplified by the table on the following page which shows the average consumption of water in Israel and Palestine.

Israeli settlements receive continuous water supply, largely from wells in Palestine, and are provided service of greater quantity per capita than that received by Palestinians in the West Bank and Gaza. When the low monthly quota levels for Palestinian municipalities and towns are approached, the remaining supply is constricted, and communities may be without water for extended periods of time. Heavy fines are imposed by the Israeli Civil Administration for pumping beyond low quota levels.

Water Demand

Current demands, however, for agricultural water exceed supplies. In the future, if the present situation and military occupation continues, no increase of the water supply will be expected to take place, except for agriculture as there is a possibility of using small quantities of treated wastewater from Palestinian cities. The Palestinian water demands per capita are expected to reach those of Israelis by the year

2020 if the peace agreements are reached between the Palestinians and Israelis. It is predicated that by the year 2010 the total amount of water needed for domestic, agriculture, and industrial purposes will be between 650-730 MCM of fresh water.

Water and the Oslo Agreement

It is now almost nine years since the initial peace conference at Madrid. Upon Israel's insistence, the peace process was divided into two tracks namely the bilateral negotiations and the multilateral talks. The bilateral were intended to lead to peace treaties between Israel on the one hand and each of the regional parties, namely, Jordan, Lebanon, Palestine and Syria on the other. The multilateral track was intended to complement and support the bilateral track by promoting regional cooperation. A special working group was established for water resources in the multilateral negotiations.

So far, a peace treaty has been reached between Israel and Jordan in which the water dispute between the two states was resolved based on mutual recognition of the "rightful allocations" of both parties to the Jordan and Yarmouk Rivers as well as the Araba ground waters. The Agreement allows for the use of Lake Tiberias for strong Jordanian surplus rain flows from the Yarmouk to be redrawn during the summer.

It also maintained the right of Israeli farmers to draw water from the Nubian sandstone aquifers from the Jordanian territory in the Araba Valley. Israel and Jordan are now working on constructing two damns in the lower Jordan River basin. There is no doubt that this bilateral agreement will not be a substitute for an integrated and comprehensive agreement among all riparians to the Jordan River basin. On the Israeli-Palestinian track, water was one of the major sticking points in the negotiations leading to the signing of the Interim Agreement (Oslo II) in Washington in September 1995. Water is referred to under Article 40 of Annex 3 "Protocol concerning Civil Affairs". The first principle in the article dealing with water and sewage states, "Israel recognizes the Palestinian water rights in the West Bank. These will be negotiated in the permanent status negotiations and settled in the Permanent Status Agreement relating to the various water resources." There is no doubt that this may be considered as an important breakthrough as it is the first time that Israel has recognize the Palestinian water rights. While the Agreement did not go into the details of the Palestinian water rights, the use of the term "various water resources" in the second sentence is very significant.

While this recognition is a very important step forward, the second and third principles in the Agreement attempt to undermine the significance of this issue by talking about maintaining existing utilization and recognizing the necessity of developing new resources, tacitly accepting that more water is needed to satisfy the needs of both populations. The Agreement states that "all powers currently held by the civil administration and military government relating to water and sewage will be transferred to the Palestinians except for those specified as issues for the "final status negotiations."

Nevertheless, the Israeli authorities have not transferred the authority for the West Bank Water Department to the Palestinian Water Authority until now. In Article 40 of the Oslo II agreement, it was agreed that the future needs of Palestinians in the West Bank are between 70-80 MCM/year of fresh water. It was also agreed that the immediate need of the Palestinians for domestic water use during the interim period is 28.6 MCM/year. The Palestinian responsibility is to supply 1.1 MCM/yr of water through the drilling of new wells, whereas the remaining 9.5 MCM/yr is to be supplied by Israel (Oslo II, 1995).

In order to honor the Palestinian commitment of providing 19.1 MCM/yr or newly supplied water resources, coordination was necessary within the framework of the joint Israeli-Palestinian-American Committee agreed upon by the Joint Water Committee (JWC) on water production and development-related projects. A project was initiated which was to being execution in July 1997. Six monitoring wells, between 300 and 700 m in depth, and six pairs of water supply wells, between 350 and 850 m in depth, were to be constructed at locations in the Heron, Bethlehem, Janin, Nablus, and Ramallah areas. Within the framework of JWC, the Israelis had given the Palestinian proposed locations, but not permission for 11 well sites for constructing Palestinian wells. These proposed wells were to be located in Hebron, Bethlehem and East Jerusalem to tap the Eastern Aquifer System. According to the working plan of the project, the wells were to be completely constructed within 18 months of initiation. So far, only two permissions have been given to construct two wells in the Hebron-Bethlehem area in the Herodion well field. Moreover, and in violation of the agreement by the Israelis, Israel has only supplied an additional 7 MCM of water per year of the 80 million to which it has committed itself. The new wells dug so far are:

- Batn el Ghul-Well No. 5 in Bethlehem district (active since 1994)

- Bala'a well in Tulkarm district (active since 1995)
- Ayn Sinya well near Jifna in Ramallah district, dug by the Jerusalem Water Undertaking (JWU) in cooperation with GTZ in 1994, but it has failed to produce water.
- Mekorot constructed a new well for the municipality of Jenin in 1996 in order to provide its population with an additional 1.4 MCM/yr as stated in the Oslo II Interim Agreement. The well has failed to extract water and negotiations are still in process between the Israelis and Palestinians as to the causes of failure in the possibility of other alternatives for water supply there.
- Hebron Municipality, in cooperation with GTZ, is currently constructing two wells for domestic purposes in the Wadi Sa'ir area; these are expected to be in operation soon.
- A new well was constructed by Mekorot in February 1996 in accordance with the Oslo II agreement for the interim period; however, it was considered by Israel as non-feasible and the well was shut down. Negotiations are continuing between the Palestinian Water Authority (PWA) and Israel regarding this well.
- Another well was constructed by Jerusalem Water Undertaking in cooperation with GTZ in 1994 at Ayn Sinya of Ramallah district but it has also failed to pump water.
- Another two wells are being currently constructed in Wadi Sa'ir area for the Hebron Municipality in cooperating with GTZ.

Regarding the overall additional 70-80 MCM/a, the bulk of this volume was to come from the eastern aquifer, where the main proportion of the water not yet exploited was brackish. Harnessing such water requires relatively large initial outlays and can pose an environmental hazard because of potential brine leakage into the source aquifer. In sum, Palestinians are getting an additional 7 MCM of water per year of the 80 million to which Israel has committed itself. So far, the Palestinians in the West Bank and Gaza have not seen the translation of this Agreement to water in their taps, but continue to experience severe water shortages. The water issue has been a contentious one, involving conflict with the United States, because Israel refused to allow the Palestinians to drill three wells in the area of Herodion. The US allocated 46 million dollars for the project which was to be carried out by an American company. Some time ago, the ministry gave permission for two of the wells, but did not grant a permit for the third — and largest-well to be drilled near the Jewish settlement of Takoa. American experts say it makes no sense to drill for the two smaller wells if they cannot drill the third well because they need all three geographically together to determine sources.

Israel intends to hold large areas of the West Bank in order to create "security zones" and to ensure that Israel's water resources are not exposed to dangers. Minister Sharon was quoted as saying: "My view of Judea and Samaria is well known, the absolute necessity of protecting our water in this region is central to our security. It is a non-negotiable item." (Boston Sunday Globe, 18 October 1998). In one of his meeting with the Palestinian negotiators, the Israeli water commission Ben-Meir noted, "I recognize needs, not rights. We are prepared to connect Arab villages to Israel as well, but I want to retain sovereignty on hand." Such statements confirm Palestinian fears of a dry peace and of Israel's genuine aspirations for peace.

Prerequisites for a Sustainable Water Agreement Between Israel and Palestine

Distinction Between Hydrology and Hydromythology

The water issue has been exploited by many Israeli politicians to serve their own agendas. Water becomes a security issue, implying that Israel is a water scarce county whose viability depends on retaining all the water resources it now controls. Security is perhaps the central concept in Israeli political dialogue — the slogan "national security" is frequently reformulated in terms of "environmental security," "food security," "water security." As de Shalis and Talis (1994) observe, the Israeli political agenda is overburdened with security issues: "Almost any political question in Israel is overridden by even the smallest security consideration." It is this obsession with security that informs many of Israel's approaches towards solving the water crisis. Above all, Israel has felt a need to have military or political control over its water supplies, and has therefore resisted perceiving the problem in terms of water rights, or in the economic terms of supply and demand, surplus and deficit. While one questions the wisdom of needing to allocate 100 cm of water per capita per year for domestic purposes, even with such a figure, Israel and Palestine have between them enough fresh water resources to meet the needs of an overall population of 21.3 million persons. When both countries reach that stage in 30-40 years, water desalination or mining of fossil aquifers can be sought.

Abandon Fantasies and Quick Fixes

Israel's proposed solutions to the water conflict have focused on "enlarging the pie" by increasing the water supplies to the region. A wide array of proposals have been made ranging from multi-billion dollar Red-Dead or Med-Dean canals, "peace pipelines" from Turkey, Lebanon, or Egypt to Medusa Bags, ferrying water from countries with water surplus to those in short supply, to tugging icebergs from northern areas, to mega-desalination projects. The recent Israeli proposal submitted to donors is to build a mega-desalination plant in Gaza to provide 50 MCM per annum of desalinated water to solve the water crisis in Gaza. The estimated capital costs of such a plant are 250 million US dollars and the estimated cost for producing each CM of desalinated water is one dollar. This means that Gazans will be spending 5 percent of their GNP to satisfy their domestic water needs. Certainly, it makes more sense to have such desalination plants in Israel, which has a larger Mediterranean shore and where the GNP is $17,000. Frankly, such dream-solutions flounder in the face of astronomical capital expenditure and environmental concerns.

Focusing on Endogenous Ways for Enhancing Supplies

Internal supply enhancement projects are economically, politically, and environmentally more feasible than the much vaunted mega-projects. These could be easily developed in both Israel and Palestine. For instance, rooftop rainwater harvesting is currently utilized in 34 percent of Palestinians houses, supplying at least 10 MCM of fresh water for domestic use. This explains how Palestinians are coping with the Israeli suppressed water supply: this simple measure could potentially provide an additional 17-25 MCM per year in the West Bank alone. The collection of rain water run-off from agricultural plastic sheeting in green houses could enhance water supplies by a further 4 MCM. Such practices would not lead to significant aquifer depletion; 75 percent of rainfall, it should be noted is lost through evaporation.

Another important water resource is the treated wastewater that could be used for irrigation. Sewage collection networks in the West Bank and Gaza Strip cover approximately 25 percent and 30 percent of the population. With the exception of the one for Bethlehem and its neighboring towns and refugee camps, most of the existing systems are old and poorly designed. Collected sewage is either discharged into open areas and valleys, such as in Nablus, Bethlehem, and Hebron, or directed towards treatment plants as the case of Ramallah, Jenin,

Tulkarm, Gaza City, Rafah, and Jabalia. A large percentage of wastewater is still collected in cesspits and open channels. Vacuum tankers are used to empty the cesspits when they become full. These tankers are owned by either the municipalities or privately. The collected wastewater is disposed of at any available location, whether open areas, streets or wadis. The few treatment plants that do exist are for the most part not functional. Over the past four years, the Palestini an National Authority has directed its efforts toward wastewater collection and treatment. Unfortunately, its efforts are being hampered by Israel's settlement policies. Most of the wastewater generated in the Israeli settlements is disposed of on Palestinian lands. Israel is stalling the process of licensing Palestinian wastewater treatment plans insisting that Jewish settlements be included. A clear case is that of Salfit, whether work on constructing wastewater collection and treatment has been stalled unless the settlement of Ariel is linked. Palestinians recognize the environmental damage caused by irresponsible wastewater dumping, but they cannot accept that this issue be exploited to legitimize the settlement policy.

These internal supply enhancement practices should be complemented by an increased focus on conservation. According to Palestinian water authorities, as much as 50 percent of domestic water is lost owing to old, inefficient supply systems. During the Israeli occupation period, the so-called civil administration invested very little in developing the Palestinian infrastructure. The Palestinian Water Authority is investing heavily in rehabilitating the water supply networks in Palestine as well as linking the 25 percent of the remaining Palestinian communities with water networks. The artificial recharge of aquifers could help to counter overexploitation of groundwater resources. Additionally, cloud seeding, a process in which chemical condensation nuclei are introduced into cloud systems, could increase precipitation by 10-20 percent (Schiller, 1993).

Mutual Recognition

Israelis and Palestinians should deal with each other as peace partners and neighbours. The suppressed water supply needs to be lifted. An important confidence building measure that Israel needs to initiate immediately is the approval to link the remaining 25 percent of Palestinian villages with piped water and to increase the freshwater supply to Gaza from its national water carrier. Even inside Israel, there is a need to do some restructuring. It is unfortunate that the Arabs in Israel who comprise 20 percent of the population receive less than two percent of the water.

- Removing the historic mistrust between the parties
- Interdependence between states is becoming the norm in world relations
- Believing that basin wide management is the ultimate goal for regional water resources.
- Realizing the unilateral steps are detrimental
- Recognition that sustainable peace should be based on justice.
- Accepting the need to live in harmony with nature. There needs to be recognition that the Middle East is an arid and semi-arid region, and that water use should be appropriate to this natural fact. Cultivated land should not be extensively irrigated; and water should certainly not be subsidized.
- Joint responsibility for the protection of water resources.

Protection of water resources should be a joint responsibility of Palestinians and Israelis. Regulatory measures for controlling the sources of contamination should be applied to the performance, technical aspects, or best management criteria for the source. Performance regulations may reflect the level of the wastewater treatment before the disposal. Technical regulations give a picture to how the source ought to be managed, operated, or maintained.

Toward a Solution

The water sector in the West Bank and Gaza Strip is one of the most important strategic sectors, which has remained undeveloped over the past three decades. The activities of the Palestinian Water Authority have been restricted by Israel to water supply administration, including operation and maintenance. Under these circumstances, the Palestinian water supply, both quantitatively and qualitatively, is in a particularly critical condition. With the signing of the Declaration of Principles (DOP) and the Oslo Agreements, Israelis and Palestinians agreed on two main issues: the equitable utilization of water resources and the joint management of the these resources. They also agreed on the development of additional water for various uses (Taba Agreement, Article 40.2). Both sides identified the two issues as principles without defining them. The translation of these principles into actual shares should be negotiated. Equitable utilization forms the basis for the allocation of the existing water resources and it is generally accepted by the International Water Law. However, the term "equitable utilization" cannot be easily quantified. In order to achieve equitable utilization of water resources, Palestinian water rights, recharge area, natural flow and population must be taken into consideration.

Palestinian water rights are summarized as follows:

- Absolute sovereignty over all the Eastern Aquifer water resources, as this aquifer is entirely located beneath the West Bank and is not a shared water resource.
- Equitable water rights in the western and northeastern aquifers, as these aquifers are recharged almost entirely from the West Bank.
- Equitable water rights in the Jordan River System: as a downstream riparian nation to the Jordan River System, Palestine is legally entitled an equitable share of the system's water resources. In this context, the Johnston Plan for Middle East water allocation, which was developed in the mid-1950s, called for, among other things, a West Ghur canal to supply the West Bank with 120 MCM to meet the needs of Palestinians. While the plan of the West Ghur canal was never implemented because of the political conflict, the Palestinian water rights in the Jordan River System are and should remain.
- Water and fishing rights in the Lake Tiberias: this natural reservoir is an integral part of the Jordan River system, in which Palestine is legally a riparian nation with the privilege equitably to utilize all of its available resources.
- Equitable rights in the Mediterranean Sea: Palestine is one of the coastal countries to the Mediterranean Sea and thus should enjoy full rights in its resources, including fishing and sailing, and should have the right to protect it from transboundary pollution.
- Full compensation for damages to Palestine's water resources caused by Israel and reimbursement for water that has been utilized by Israel during the occupation.

On the other hand, the annual safe yield of the aquifers is apportioned according to the extent of recharge area. About 80 percent of the recharge area of the western basin is located within the West Bank while only three percent of the Jordan River's basin fall within Israel's pre-1967 boundaries. Eastern groundwater basin is an unshared groundwater basin as both recharge and storage areas are located within the boundaries of the West Bank. If it is accepted that allocation of water rights would be made according to equal per capita shares, the total quota of each side would be proportional to the population size. Thus, the 2086 MCM of water available within mandate Palestine would be shared so that Palestinians get 698 MCM instead of 238 MCM which is currently used. The Israeli share should be 1388 MCM instead of 1959 which is currently consumed by the Israelis. The above distribution of water rights between the two sides is building on the population figures. The per capita consumption for both the Palestinians and Israelis will be 241 [m]/a.

Agreement on the Allocation of Available Water Resources

Resolving the water conflict between Israelis and Palestinians as outlined here is of paramount importance. First, it will introduce for the first time in the region, an integrated water management scheme if adopted and will certainly be of great value for resolving the water conflicts between Syria, Lebanon, Israel, Jordan and Palestine. Second, it will show the opponents of the peace process that negotiations are possible. For the politicians, it would lessen the chances of conflict; for industrialists and agriculturists, it would foster stable growth; for every citizen, it would result in guaranteed regular supplies of household water.

The resolution of the Palestinian-Israeli allocation and water rights disputes need to be governed by the principles of international law. Two legal aspects of the conflict are of concern. First, Palestinians and Israelis must reach a consensus on sovereignty over water resources in the West Bank and Gaza. Second, Palestinians and Israelis must reach agreement on rightful allocation of shared water resources to each party. Negotiations over allocations and water rights should be conducted with an eye on justice rather than might, and independent arbitration may be necessary. The international community and financial institutions should be asked to make clear to all parties that loans for international waterway projects will not be forthcoming until the agreement is negotiated.

The Krasnoyarsk Krai in Sustainable Water Management Indices

According to the prognosis of Rodda (1997), between 2035 and 2045, fresh water consumption in the world will equal accessible water resources. Following the 'petroleum' period in Russia, water resources may provide an advantage for Russia in the future (Danilov-Danilian 2007) because Russia has 10% of the world river runoff and 20% of non-freezing fresh water in Lake Baikal. The complexity and variety of modem water management problems in Russia has led to the need to develop a systems approach and apply programming methods to solve the existing problems, and realise complex consecutive measures directed to the development and perfection of a system of water resource management that provides protection from pollution, and supplies water for the economy and population in terms of both quantity and quality.

Although Russia has decreased the volume of water intake and sewage water discharge as a result of an industrial production recession, no visible improvement in surface water quality has been observed. On average, 21% of Russian water samples do not satisfy the requirements of the State Standard Drinking Water regulations. Most rivers and lakes in Russia are contaminated by waste from human and economic activities: the quality of surface water practically everywhere does not satisfy sanitary hygiene requirements. Wasteful use of water resources is on-going. These problems are aggravated by extremely low financing of water supply and water protection arrangements, and the many legal nature protection standards that often contradict one another (Anon. 2000).

Government policy in the treatment, recycling and protection of water was affirmed by the board of the Russian Federation Ministry of Natural Resources (proceedings from 23.06.99 No. 10/1), the guiding document for definition of the normative and legal basis, development of organisational and economic mechanisms, and planning and carrying out technical activities in relation to the territory of the Russian Federation. This document states that the 'strategic aim of water policy is sustainable water management', and 'aims and interests of state water policy are expressed in long-term principal indices of water objects' (Anon. 2000). Nevertheless, such indices for Russia have not been formulated.

Indicators of sustainable water management, suggested by the UN (1996), do not contain elements such as maintenance of natural water reproduction, maintenance of sanitary-epidemiological water state, conservation of biological water diversity, creation of socio-economic water consumption functions, or instruments of water policy to conserve sustainable management of water consumption and protection from injurious water effects. Some preliminary approaches to the problem have been presented in our previous papers (Shaparev 2005, 2007). In this chapter, the authors have attempted to produce a complex system of indices, which take into account both known suggestions and the authors' point of view on a solution for water management problem, by combining the UN indicators with the indicators from statistical report 2-TP, affirmed in Russia (Decree of Goskomstat of the Russian Federation 2000), and extended by the authors. The complex system of sustainable water management indices is subdivided into criteria and indicators, analogous to the system for sustainable management of Russian Federation forests. Criteria are the main direction for practical activity to satisfy the aims of state policy on observance of general principles, demands and mechanisms to realise sustainable water management. Criteria are realised and assessed from the sum of the characteristic indicators. Indicators are quantitative and qualitative characteristics of criteria: temporary state of water objects, their production, presence of basic norms, protection of the population from injurious water effects, etc. (Liu and Chen 2006). The system of criteria and indicators provides an opportunity to assess directions of changes in water management.

The River Yenisei basin, in the centre of Russia, covers 78% of the Krasnoyarsk Krai. The economic activity in the Krai represents activity in the Yenisei basin, and a system of sustainable water management indices was applied to this basin. To present the indices, we used government report data, and scientific papers and monographs. However, because of a lack of data, a part of the indices suggested by us cannot be incorporated and characterised.

Indices of Sustainable Water Management for Yenisei River Basin

Natural Water Production

The main aim of this criterion is maintenance of natural water reproduction. The Yenisei River basin covers 2.58 million km^2, and 78% of the Krasnoyarsk Krai (including Taimyr and Evenkiya), with 1.84 million km^2 belonging to the basin, and the remaining areas mainly belonging to the basins of the rivers Ob', Lena, Khatanga and Pyasina. The volume of the Yenisei River runoff is 600-630 km^3 per year, 14% of the annual runoff in Russia and 1.475% of the world annual runoff (Shaparev 2002). The total volume of groundwater in the Krai is 9-10 km^3 per year, with 25% intake (Korpachev and Babkina 2001). The total area of the Yenisei basin under exploitation at present is about 28% (the Krasnoyarsk Krai without Taimyr and Evenkiya). Natural gas and oil prospecting and extraction in these territories will expand the exploited area in the future.

The ratio of permissible (planned, calculated) and real water intake in the basin consists of three components: the volume of water intake that does not affect the natural hydrological regime, natural purifying ability and biological productivity. The description of this indicator is aggregated.

The annual sum of rainfall in the Krasnoyarsk Krai increases from north to soutihover about 3000 km, from 200 to 1200 mm, and evaporation similarly increases from 200 to 400 mm (Chekha and Shaparev 2004). The average annual balance of water production in the Yenisei basin is, therefore, rainfall-1300 km^3, evaporation-550 km^3, river runoff-600 km^3.

The area of cultivated land in the Krasnoyarsk Krai was 2,977,700 ha or 4.1% of the Krai area in 2005 (with total area of 72 million ha) (Government Report 2007). Between 1991-2005 in the Krai, 19,687 ha of land were affected, further landimprovement activities affected 19,779 ha and 19,138 ha were finally cultivated. From 1993 onwards, there was a reduction in the annual area of affected land and extension of areas of cultivated lands was observed (Government Report 2007).

According to data (Raspopin 2005), about 40,000 ha of forest is cut each year and forest is replacing agricultural cultivation because of a decline in land productivity, exclusion of land from agricultural use and overgrown areas covered in forests and shrubs. According to data (Raspopin 2005), the annual area of reforestation in the Krai is about 40,000 ha, similar to the area that is cut, but the proportion of coniferous forest is decreasing while deciduous forest is increasing. At the same time, the area of recultivated forestland is 14,971 ha (78.1% of all recultivated area in the Krai). Thus, in 2005, forest resources in the Krasnoyarsk Krai were extended by 5900 ha compared to values in 2004. The area of forest in the Krasnoyarsk Krai (without Taimyr and Evenkiya) in 2003 was 57.9 million ha (Shaparev 2002), equal to 80% of the basin area.

Maintenance of the sanitary-epidemiologic State of Surface Waters

The main aim is to maintain water constant and the planned aim is to decrease injurious effects and improve and maintain water quality. The water quality of most of the Krai does not satisfy the standards. The main source of contaminants is sewage water from industry, agriculture, utilities and surface runoff. Surface waters in the Krai are contaminated almost everywhere with mineral oils, phenols, copper, zinc, iron, aluminum, manganese and arsenic, and are classified as 'dirty' or 'very dirty'. The sanitary-chemical and the sanitary-bacteriological indices of open water sources in the Krasnoyarsk Krai have worsened; the former being below the average in Russia and the latter being higher. Heavy metals (mercury, lead, cadmium), but no infectious agents, were not found. The quality of groundwater is also unsatisfactory, with increasing chemical contamination, although the bacteriological indices improved by 2000, but are now worsening.

The poor sanitary-hygienic state of drinking water sources is caused by ineffective disinfection facilities, absence of properly organised zones of sanitary protection, insufficient control by management of their territories, and natural contamination of water in auriferous strata. Furthermore, the drinking water pipelines in 243 localities of the Krasnoyarsk Krai, with a total population of 351,840, are epidemiologically insecure. Most are situated in the Shushenskii, Yeniseiskii, Sharypovskii and Turukhanskii districts. In the Idrinskii, Motyginskii and Novoselovskii districts, 50-85% of the samples did not satisfy sanitary stan-dards for bacteriological indices. The standard water supply for 21 rural localities in these districts, with a population of 21,709, is epidemiologically dangerous. There are no data on coliform content in freshwater in government reports, although they were assessed in drinking water from the city of Krasnoyarsk in 2002 (Babkina and Koren'kov 2003) where 44 water samples were taken from centralised water supply sources, and 49 samples taken direcdy from the distribution point. Only four samples, taken in a factory, were unsatisfactory in terms of the sanitary norms. No pathogenic and likely pathogenic organisms were found in the drinking water in 2002.

Water from sources in Komarovo village (Kanskii district) is unacceptable for drinking because of high levels of alpha and beta radionuclide isotopes of uranium238. 30% of the investigated water samples from artesian sources in the Krai have a high content of uranium derivatives. Also, total beta-activity in the centralised water supply sources in the Krasnoyarsk Krai did not reach the acceptable value in 2005, while the total alpha-activity exceeded acceptable levels by 2.5-times in 39 of 109 samples (Government Report 2007). Water in the Yenisei River downstream of Zheleznogrsk had higher radioactivity but, according to data (Government Report 2005) 100 to 1500 km downstream from a sewage water pipe, the volumetric activity of strontium 90 and caesium 137 is practically at background levels. With a total discharge volume of all contaminants in 2005 of 687,900 tons (point sources: 74.5%, auto transport: 25.5%), and the Krai area (without Taimyr and Evenkiya) covering 72 million ha, the average air contaminants total 0.95 tons km^{-2}.

With decreasing discharges into the Angara River in 1998 on the borders of the Krasnoyarsk Krai, the total volume of contaminated water in the river increased 14-fold because of output from Irkutskaya Oblast', increasing to 8.4 million m^3 of sewage water on the borders of the Krai, and 861 million m^3 on the borders of the Oblast'. The natural hydrological regime of the Yenisei River has been disturbed by the Sayano-Shusheskaya, Maina and Krasnoyarsk hydroelectric power stations. Water from the Krasnoyarsk reservoir flows to tail-water, at a constant temperature of 4°C, which results in the formation of ice-holes up to 200 km downstream in winter, and affects recreational zones in Krasnoyarsk. The total discharge of Krasnoyarsk hydroelectric power station is 80 km^3 per year, and the discharge of the SayanoSushenskaya and Maina power stations is about 40 km^3 (Shaparev 2002). According to the data for 1994 (The State of Environment 1995), forest resources in the Krasnoyarsk Krai (including Taimyr and Evenkiya) were 89.7 million ha, and 5.21 million ha (5.8% of these have a primary water protection function). There were 57.9 million ha of forest resources in the Krai (without Taimyr and Evenkiya) in 2005 (Government Report 2007), with water-protecting forests constituting approximately 8.9%.

According to the RSFSR Council of Ministers Decree No. 384 from 25.09.87 on stopping timber rafting on rivers and other water sources, this should have stopped in the Krai in 1994. On the Taseeva River, this practice was reduced to a timber volume of 189,000 m^3 from the 1993 value of 287,000 m^3. Timber rafting with ships was carried out in 1994 on six rivers: Angara, Yenisei, Kas, Sym, Kazyr, Tuba (State of Environment 1995), and still takes place.

The measurement unit of water discharge is the percentage of sewage water that is subjected to adequate purification before discharge. Purified water should not harm the watercourses or the environment. The discharge of insufficiently purified water worsens the quality of water resources, which affects population health and leads to economic, social and ecological instability. The structure of sewage water discharge in 2005 was, by type of economic activities, as follows: municipal economy-78.9%, transport and agriculture-0.7%, industry-20.3%; by the branches of industries: energy sector-12.0%, chemistry and petroleum chemistry-9.0%, non-ferrous metallurgy-4.7%, woodworking-36.0%, coal industry 35.6%, and other-2.8%.

From 1993 to 1998 the total volume of discharge in the Krasnoyarsk Krai reduced from 2.806 km^3 to 2.393 km^3 (14.7%), clean water (without purifying) from 1.997 to 1.763 km^3 (11.7%); not sufficiently purified water from 0.584 km^3 to 0.492 km^3 (15.8%); polluted (without purifying) water from 0.198 km^3 to 0.141 km^3 (28.8%); and nominally purified water from 0.03 km^3 to 0.024 km^3 (20%). In Russia the indices from 1993 to 1998 were as follows (Demin 2000): total discharge reduced from 68.2 km^3 to 55.7 km^3 (18.3%), nominally clean discharge from 38.4 to 31.2 km^3 (18.7%), not sufficiently purified water from 18.7 to 15.8 (15.5%), polluted (widiout purifying) water from 8.5 to 6.2 km^3 (17%), and normatively clean water 2.6 to 2.5 km^3 (4%). In the structure of sewage waters in the Krasnoyarsk Krai for the above period, the specific share of polluted water (insufficiendy purified and discharged widiout purification) dropped from 27.9% to 26.4%. The share of normatively clean water also reduced from 1.069% to 1.0029%. In Russia as a whole the first index reduced from 39.9% to 39.5%.

Protection from Injurious Water Effects

The main aim is protection of the population and economy from injurious effects of floods, waterlogging, water erosion, droughts, etc. This is made up of indicators for flood protection, flood protection engineering, and erosion prevention in the catchment area.

Development of a System for Providing Quality and Quantity of Water for the Population

The main aim is creation of conditions for uninterrupted supply of economic and drinking needs for all within sanitary hygienic norms. The number of people using drinking water containing high amounts of compounds of the 2nd danger class (fluorides, strontium, benzol, barium, chloroform, tetrachlorethylene) was 183,000. Water containing dangerous concentrations of compounds of the 3rd danger class (iron, manganese, nitrates) is used for drinking by 725,000 people. Thus, 30.3% of the Krai's population in 1992-1993 was supplied with water of unsatisfactory quality by hygienic norms and chemical indices. More than 227,000 people drink water with a high content of iron, fluoride, manganese, nitrates, ammonia, sulphates, hydrogen sulphide, and benzpyrene; about 90,000 people are subjected to water with high radioactivity; and 860,000 citizens in Krasnoyarsk occasionally drink water with a high chloroform and phenol content. The centralised water supply in the Krai covered 85.57% of the population (64.0% urban, 21.57% rural) in 2001. The non-centralised water sources (tube and dug wells, springs) supply 13.94% of the population, and 0.48% of the population use water taken directly from the river. Although the Krai has a relatively high provision of centralised watersupply systems, the state of these systems is unsatisfactory: one in five water intake points do not have a sanitary protection zone; one in seven does not have disinfecting facilities; and one in ten does not have water treatment facilities. The rate of basic asset wear and tear amounts to 34%, with about 120 km of water pipes annually needing to be replaced. The water pipes that did not satisfy sanitary norms increased from 1994 to 1998 from 29.8% to 3.5%. It is important to note that the state of the water drainage system is also unsatisfactory. The rate of wear and tear of the sewage pumping stations and sewage purifying systems is 80-100%. This requires renewal of 170-180 km of sewage pipes annually, total reconstruction of the present sewage pumping stations and sewage purifying systems, improvement in sewage water post-treatment facilities and reduction in the mass of pollutants, development of facilities for sediment treatment and utilisation, and introduction of disinfection facilities (Mochalov 2001). The underground economic and drinking water supply points in the Krai (wells, springs) are mosdy non-affirmed and not passed by the state, due to poor hydrogeological studies of the area and a direct contradiction of article 29 of the Federal Law on subsoil. To meet the needs for water in 2005, 13.2% came from groundwater (in Russia in 1997 centralised water supply utilised 35% groundwater (Demin 2000)). 77% of the groundwater intake is used in the economy (89.5% economic and drinking water supply). The volume of groundwater intake in the Krai is 1,268,600 m^3 per day or 0.465 km^3 per year, taking into account that the predicted exploitation of groundwater resources, including riverside infiltration water supply points, is about 10 km^3 per year or 4.65%.

Surface water extraction in 2005 accounted for 86.8% (including water circulating in river valleys in alluvial deposits and hydraulically connected with surface water). In Russia in 1997 sources of centralised water supply were provided from 65% surface water (Demin 2000). The total water intake in 2005 in the Krasnoyarsk Krai (without Noril'sk industrial district) amounts to 2731 million m^3. A total of 2508.4 million m^3 were used for industrial needs, and 2424 million m^3 were discharged. Analysis of recent data shows that the anthropogenic load on water in the Yenisei River basin has reduced, but it has increased in the Chulym River basin, while diere has been no change in the Angara River basin. The main water intake is from the Yenisei and Chulym, so the maximum sewage water discharge occurs in theses basins.

The volume of annual water intake is 2.7 km^3, while the annual river runoff is 600 kmsup 3, so usage amounts to 0.45%. Considering that the central and southern districts are the main areas of water consumption and only 20% is runoff, the use of surface waters is 2.2%. The surface water losses through transportation and from pumping into aquifers and storage and discharge sewage waters in 2005 amounted to 69.8 million m^3 (2.5%). In Russia this index was 8.6% (Demin 2000). Also 23% of all intake groundwater is lost during transportation and discharge, without including mine spoil drainage.

On average in the Krai, specific water consumption amounts to 243 l per day per capita (Mochalov 2000). From the level of providing the population with drinking water, cities and towns in the Krai can be subdivided into three groups. Measures are needed to install additional sources in cities and towns of the first group, and a program for reducing water consumption in cities and towns of the third group. According to the data (Demin 2000), the specific water consumption for economic and drinking needs in the East Siberian economic region increased from 1 970-1 998 by 48 l per day, or 330 l per day per capita. On average in Russia these figures were 1181 and 350 l in Moscow or 198 l and 610 l per day per capita. Other indicators are consideration of the cost of distributing 1 m^3 of water for economic/drinking needs; development and introduction of new effective technologies for purification of surface, ground and sewage waters; and development and introduction of new technologies for disinfection of drinking and sewage waters.

The density of the hydrological network is the average area served by one hydrological station. Note that the density of hydrological networks should be sufficient to provide the necessary information for water resource assessment, development and management. The hydrological network density is determined by the economic state of a country, the population density, climatic features and geographic zones. The government network for observing water state in the Krai (according to data of 2001) includes 137 hydrological stations. The average area served by one station is over 17,000 km^2, which is clearly insufficient for the Krasnoyarsk Krai. It has been suggested that a complex system of sustainable water management indices for a basin might be used in the Russian Federation to create a common network for ecological water monitoring; in development of regional programs on development of water supply complexes; in project development of water protection zones and schemes of complex use and protection of water; in creation of a water cadastre using geoinformation technologies; for development of regulations to limit permissible injurious effects on water and limit permissible discharge to river basins to determine the level of economic safety of using water resources; and finally to contribute to sustainable nature management in the basin. The developed system of indices has been applied to analyse the state of the River Yenisei that borders the Krasnoyarsk Krai. Even results from incomplete indices analysis give grounds to consider the situation in the Krasnoyarsk Krai as, at best, unsatisfactory. In spite of positive elements, such as the considerable amount of water-protecting forests, land-improving activities, the absence of biological water pollution, and the absolute prohibition of timber rafting, the negative phenomena are still very important. A considerable part of the population in the Krai is affected by noxious substances in the water, which endanger health. Over a number of years with a reduction in total discharged volume, the share of normatively purified waters has not increased. There is an insufficient number of stations for observing the water qualities. It should also be noted that biological resources of surface waters are not adequately used.

To implement sustainable water management (water resources should be inexhaustible) in future, there should be a decrease in freshwater intake for industrial needs; no disproportionate freshwater intake with increased industrial production; decreased drinking water consumption; and development of a circulating water supply by industry. Preservation and improvement of water resources in the Krasnoyarsk Krai must be achieved by improving purification facilities and water supply systems in inhabited localities, and delimiting water protection zones, especially in industrial centres. It should also be noted that sustainable water management cannot occur without directing public opinion and obtaining conscious participation of the population in the solution of sustainable water management problems in regions of Russia.

The water quality of the Yenisei River according to the index of water pollution (IWP) is characterised as 'polluted' (Demin 2000), with some sections ranging from 'moderately polluted' to 'dirty'. The IWP for branches of the Yenisei (Kacha, Kan, Angara) and Chulym Rivers can be characterized in some sections as 'polluted' or 'dirty'. The Krasnoyarsk Krai also has a higher volume of sewage water containing pollutants than the other 16 regions in Siberia. At present the drinking water for the population is not improving because of insufficient water protection activity and lack of financing from regional and federal budgets. It is important to emphasise that the surface waters do not conform to the Russian water quality standards both in the Krai and in Russia as a whole. There has been no recent improvement in water quality, even though this problem requires urgent attention.

Bibliography

Aung, K. Hla, Thomas F. Scherer: *Introduction to Micro-Irrigation*, AE, NY, 2003.

Bander, J.: *Scheduling Irrigation with Evaporation Pans,* Montana State Univ. Agr. Bull. Montana, USA, 1984.

Barghouti, S. & Le Moigne, G.: *Irrigation in Sub-Saharan Africa: Development of Public and Private Systems,* World Bank, Washington, DC, 1990.

Bhakar S.R. : *Ground Water Hydrology : Theory and Practice*, Agrotech, Delhi, 2009.

Bos, M.G. & Nugteren, J.: *On irrigation Efficiencies*, Wageningen, the Netherlands, Int. Inst. Land Reclam, Improve, 1978.

Burman, R.D., Cuenca, R.H. & Weiss, A.: *Advances in Irrigation*, Orlando, Academic Press, Florida, USA, 1983.

Caneva, K. L.: *Robert Mayer and the Conservation of Energy*, Princeton University Press, Princeton, New Jersey, 1993.

Charles, M. Burt: *The Surface Irrigation Manual*, Waterman Industries, California, 1995.

Copeland, M.C.: *A Manual for Irrigation Planning in Developing Areas*, Pretoria, South Africa, SECOSAF, 1993.

Devi, Sudharmai: *Analytical Procedures in Soil Science and Agricultural Chemistry*, Agrotech, Delhi, 2004.

Dreybrodt, Wolfgang: *Processes in Karst Systems Physics, Chemistry and Geology,* Springer-Verlag, New York, 1988.

Dunai, T.J.: *Cosmogenic Nucleides,* Cambridge University Press, U.K., 2010.

Engelmann, S., & Carnine, D.: *Theory of Instruction: Principles and Applications*, ADI Press, Eugene, OR, 1991.

Fischer, Louis: *Oil Imperialism: The International Struggle for Petroleum.* New York: International Publishers, 1926.

Gilley, J.R.: *Advances in Irrigation,* Orlando, Florida, Academic Press, USA, 1983.

Goldstein, Martin, and Inge F.: *The Refrigerator and the Universe*. Harvard Univ. Press, 1993.

Hanks, R.J.: *Efficient Water Use in Crop Production*, Madison, Agron, 1980.

Hessayon, D. G. Dr.: *The Vegetable Expert*, England, PBI, Publications, 1985.

James, L.G.: *Principles of Farm Irrigation System Design*, Wiley, New York, 1988.

Kenneth H. Solomon and Greg Jorgensen: *Subsurface Drip Irrigation*, CATI Publication, US, 1993.

Mach, E.: *History and Root of the Principles of the Conservation of Energy*. Open Court Pub. Co., IL, 1872.

Macself, A.J.: *Soils and Fertilizers*, Satish Serial Pub, Delhi, 2005.

Mahajan Gautam : *Ground Water : Surveys and Investigation*, APH, Delhi, 2009.

Narasimha Rao, P.: *Irrigation Development: Issues and Challenges*, Discovery, 2007.

Orson, W. Israelsen: *Irrigation: Principles and Practices*, Axis Books, 2010.

Petr, Tomi: *Fisheries in Irrigation Systems of Arid Asia*, Daya, 2007.

Pradhan, Prachanda and Upendra Gautam: *Farmer Managed Irrigation Systems in the Changed Context*, Farmer Managed Irrigation Systems Promotion Trust, 2002.

Punmia, B.C.: *Irrigation and Water Power Engineering*, Laxmi Publications, Delhi, 2009.

Qassim, Abdi: *Sprinkler Irrigation*, Situation Analyse, IPTRID, UK, 2003.

Raju, K.S., Biere, A.W., Kanemasu, E.T. & Lee, E.S.: *Advances in Irrigation*, Academic Press, Florida, USA, 1983.

Rawitz, E. & Hillel, D.: *Drip Irrig.*, San Diego, California, USA, 1974.

Rimal Gautam, Suman: *Incorporating Groundwater Irrigation: Technology Dynamics and Conjunctive Water Management in the Nepal Terai*, Orient Blackswan, 2006.

Shalhevet, J.: *Irrigation of Field and Orchard Crops under Semi-arid Conditions,* Bet Dagan, Israel, Intl Irrig. Info. Ctr., 1976.

Thomas F. Scherer at all: *Sprinkler Irrigation Systems*, MWPS, 1999.

Vaux, H.J., Jr & Pruitt, W.O.: *Advances in Irrigation*, Orlando, Florida, USA, Academic Press, 1983.

Withers, B. & Vipond, S.: *Irrigation: Design and Practices*, Ithaca, NY, USA, Cornell University Press, 1980.

Index

❑❑❑